Microsoft® Word for Engineers

Sheryl Sorby
Department of Civil and Environmental Engineering
Michigan Technological University, Houghton, Michigan

Microsoft Word for Engineers

ADDISON-WESLEY
An imprint of Addison Wesley Longman, Inc.
Menlo Park, California · Reading, Massachusetts · Harlow, England
Berkeley, California · Don Mills, Ontario · Sydney · Bonn · Amsterdam · Tokyo · Mexico City

Senior Acquisitions Editor: T. Michael Slaughter
Developmental Editor: Judy Ziajka
Editorial Assistant: Royden Tonomura
Production: Caroline Jumper
Copyeditor: Christine Sabooni
Proofreader: Holly McLean Aldis
Indexer: Nancy Kopper
Cover Design: Yvo Riezebos
Text Design: Side by Side Studios
Composition: Craig W. Johnson & Simmy Cover

This is a module in the Engineer's Toolkit, an Addison-Wesley SELECT edition. Contact your sales representative for more information.

Photo Credits:
Chapter 1: NASA/Mark Marten/Photo Researchers, Inc.
Chapter 2: Courtesy of Intel Corporation
Chapter 3: NASA/LBJ Space Center
Chapter 4: ©Bill Bachman/Science Source/Photo Researchers, Inc.
Chapter 5: ©Erich Schrempp/Photo Researchers, Inc.
Chapter 6: Courtesy of National Center for Supercomputing Applications
Chapter 7: © Alan L. Dietrich/Photo Researchers, Inc.

Library of Congress catalog card number 95-131339.

ISBN: 0-201-35092-0

3 4 5 6 7 8 9 10—CRK—01 00 99 98

Addison Wesley Longman, Inc.
2725 Sand Hill Road
Menlo Park, CA 94025
http://www.aw.com/cseng/toolkit/

Contents

1 Engineering Design and Documentation

Communication Systems

Elementary-school children across the country participate each year in a program called Marsville. In this program, students are asked to imagine that they are the engineers and scientists in charge of the next significant program in space exploration. They must design the food, water, air, and communications systems necessary for survival on the planet Mars. Communications may not seem as important as food, air, and water, but effective communication with others is vital to survival in a hostile environment. In fact, communication skills are among the first skills you learn as a child, and the ability to communicate effectively has played an integral role in your development, education, and socialization.

Good communication is also essential to engineering. Engineers communicate their ideas and designs by three primary methods: graphical, written, and oral. The remainder of this text focuses on written communication in the engineering profession. This chapter describes the five stages of the engineering design process and the kinds of documentation that support each stage.

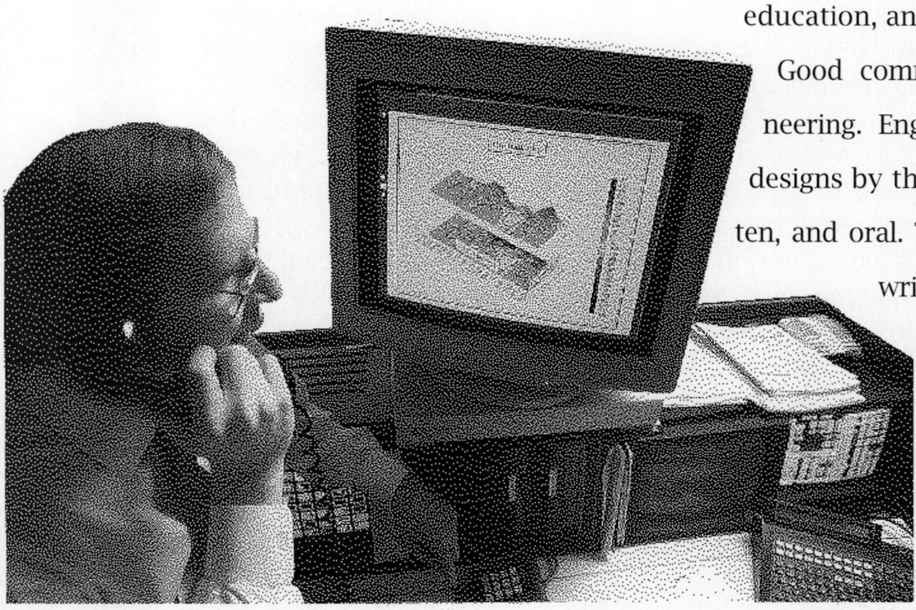

INTRODUCTION

This chapter discusses the role of problem solving in the daily life of an engineer. You learn about the five stages of the engineering design process and the types of problems that engineers must address during each stage. You also learn about the types of documents engineers typically create at each stage of the process.

1-1 PROBLEM SOLVING AND COMMUNICATION

Virtually everything that you touch in any given day has been engineered, from the clothes you wear to the car you drive to the computer you use. One or more engineers designed these products, analyzed them, determined how to mass-produce them, or designed the machines used in their production. Even much of the food you eat was produced or delivered with the help of engineers. An engineer determined how to process and package the food, which was then brought to your supermarket in a vehicle designed by engineers, which used a transportation system also designed by engineers.

Engineers focus on solving problems throughout their working careers. This focus on problem solving distinguishes engineering from other professions.

The problems engineers must solve vary greatly by discipline and degree of difficulty. Some problems are relatively simple. For example, designing the power system for a home usually is a straightforward task that is largely dictated by building codes. Other problems are complex and involve the solution of numerous smaller problems. For example, designing a bridge requires the solution of many problems, from deciding which materials to use to determining the style and capacity of the bridge.

Unlike most other types of problems, engineering design problems typically involve more than one possible solution. It is the engineer's job to weigh factors such as cost, longevity, serviceability, manufacturability, and aesthetics as well as human factors before selecting the *best* solution to the problem at hand.

For instance, when designing a bridge, the engineer must choose the materials (steel, wood, concrete, or plastics), the bridge style (cable suspension, truss support, or arches), the number of traffic lanes, whether to include a feature like a drawbridge, and whether to put in a pedestrian walkway or bicycle path.

These are just a few of the choices the engineer must make in designing the bridge. There are no perfect choices or solutions. The engineer's job is to consider the pros and cons of a number of potential solutions and to choose the one that appears to be the best.

In choosing construction materials for the bridge, the engineer must weigh the advantages and disadvantages of each and then select the materials best suited to the job. Concrete can carry heavy loads and can be formed into many different shapes, but it requires massive amounts of material, is brittle and prone to cracks, breaks down under repeated freeze-thaw cycles, and is not always aesthetically pleasing. Steel carries heavy loads very efficiently but is relatively costly, and in a damp environment, it is susceptible to corrosion. Wood is inexpensive and beautiful, but it is also prone to corrosion in a wet environment and is difficult to form into different structural shapes. A combination of all these materials may actually be used to construct a complicated structure such as a bridge.

In solving problems, engineers typically work in teams. In designing and building a bridge, one team designs the bridge supports, another team analyzes the stability of the ground under the bridge supports, and another team oversees the construction of the bridge. A project manager supervises the work of each of the teams.

Because engineers tend to work in teams to solve problems, and because product designs must often satisfy a diverse group of people, clear communication is extremely critical if the design is to result in a successful product. Engineers communicate by graphical, written, and oral means. As a student and as an engineer you will spend a significant amount of your time preparing written documents and making oral presentations.

Written documentation is especially important to engineers because it enables them to learn from the experiences of others and to build the knowledge of the profession as a whole. For example, if a design fails at some point during its life cycle, it is important for engineers to review the entire design process to determine where it went wrong to produce a more effective design in the future. Without proper written documentation, this type of analysis is not possible.

This module is designed to help you develop your writing skills and to teach you how to prepare engineering documentation using Microsoft Word 97. As you work your way through the module, you will be introduced to the kinds of documents that engineers typically prepare at each stage of the design process. You'll also learn the features of Word that you can use to create these documents.

1-2 THE ENGINEERING DESIGN PROCESS

Written documentation is often required at each stage of the engineering design process. For example, if a new project shares aspects of an already completed project, you can examine the documentation from the original project and modify it to fit the new project, often saving a good deal of time. Or if legal proceedings result from an engineering project, the written documentation can help settle disputes. But most important, documentation is essential for communicating information to design team members or to a client at particular phases of the project. Engineering design is an iterative process. When designing a bridge, the engineer first estimates the weight of the pavement and supporting structures. The bridge is then designed, and the actual weight of the bridge components can be accurately determined. Based on these calculations, the bridge supports may need to be redesigned to account for any differences between the estimated weight of the structure and the computed weight of the structure. This process may be repeated several times before the final design solution is obtained. The following table links the type of documentation that supports the work occurring in each *step* of the design process.

General Problem-Solving Procedure	Word 97 Problem-Solving Procedure
1. Define the problem	Letters and memos
2. Gather information	Experimental reports
3. Generate and evaluate potential solutions	Proposals
4. Refine and implement a solution	Technical papers
5. Verify the solution through testing	Design specifications

Step 1: Define the Problem. The first step in the engineering design process is to define the problem. During this step, you should determine your objectives. You should also consider what information you are given, what additional information you may need, and the constraints that will affect the solution to the problem. You will often use letters and memos to communicate with others about the problem definition in this step of the design process.

Step 2: Gather Information. The second step involves gathering the information you need to solve the design problem. This phase may include experimental work, computer analysis, or a review of similar work. At this step, you will frequently write experimental lab reports.

Step 3: Generate and Evaluate Potential Solutions. This step involves hypothesizing several possible solutions to the problem. Because engineering involves making choices and trade-offs, it is important to examine several alternative solutions. At this step of the design process, you might write proposals to investigate one possible solution more thoroughly.

Step 4: Refine and Implement a Solution. During this step, you evaluate all the possible solutions and choose the best one. In determining the best solution, you may weigh factors such as cost, strength, durability, and aesthetics. At this step of the design process, you may be asked to write a technical report or paper describing the selected solution and to make an oral presentation to sell the design solution to others.

Step 5: Verify the Solution Through Testing. The last step in the design process is implementing and testing the solution. In implementing the design, the product is manufactured or constructed. You must also test and evaluate the product, and you may need to modify it, based on the test results. During this step, you may be asked to write design specifications that tell the contractor or manufacturer exactly how to fabricate the product and what type of material to use.

The rest of this module teaches you about the kinds of engineering documentation you are likely to create throughout the design process. You will learn about a new type of documentation, as well as a new Word feature, at each step in the design process. Some readers may find this organization artificial. For example, at the second step of the design process (gathering the data), the Word procedure for creating tables is introduced. In reality, experimental reports often contain equations and figures in addition to tables. However, it is impractical for you to learn all that you need to know about Word by the third or fourth chapter of this module. For this reason, we introduce new Word features—which correspond to the type of documentation discussed—at a rate of roughly one per chapter. After studying this text, and trying all the Try It! Exercises and end-of-chapter problems, you will have learned the essential features of Word and will understand how to use them to create useful engineering documentation.

SUMMARY

For the design process to be carried out efficiently, it is critical that the various engineering teams involved in a project communicate their ideas and their intentions. Engineers primarily communicate with one another by drawings and documents. In this chapter, you learned about the five steps of the engineering design process and about the types of engineering documentation likely to be used at each step. In the following chapters, you will learn how to use Word to create these various types of documentation.

2 Introduction to Microsoft Word

Microprocessor

Perhaps no other industry has matched the technological strides achieved by the computer industry in this century. One theory has it that if the automotive industry had accomplished as much as the computer industry over the past 75 years, we would all be driving Rolls Royces, which would get 200 miles per gallon, and the cars would cost only $10 each. The microprocessor is perhaps the most significant achievement of the computing industry. The development of the microprocessor has led to an explosion of computer use, which has dramatically affected how we communicate. For example, word processing enhances our ability to effectively convey our ideas. This chapter explains some of the basic concepts of word processing software and introduces you to Microsoft Word.

INTRODUCTION

In this chapter, you begin to learn the basics of Microsoft Word 97. You learn how to start Word and then become acquainted with the components of the screen. Next, you find out how to create a new document. You learn how to print documents, how to save them for later use, and how to open and close existing documents. Finally, you discover how to use the online Help system to find out more about Word.

The best way to learn any new software package is to sit at the computer and experiment. When you are first starting out, you may want to use this text as a guide to help you use the software. This text includes a number of short exercises called Try It! exercises. You should work through as many of these as possible; the more of them you try, the faster you will learn the software. Do not be timid, and do not restrict yourself to attempting only the exercises given.

This module is not meant to be a comprehensive guide to Word. If you need additional help, read the Word manuals or consult the online Help system.

2-1 USING WORD PROCESSORS

In the past, written engineering documentation was created on typewriters. When personal computers became affordable in the 1980s, they gradually replaced typewriters, and word processing programs such as Microsoft Word became popular. *Word processing* is similar to typing documents. When you use a word processor, however, your document is stored in the computer. This makes it easy to edit the document and correct any mistakes before you print it.

Word processors have revolutionized the ways that we prepare documents and communicate with one another. Word processors have become so sophisticated that they include many features that required special typesetting capabilities just ten years ago. In fact, the term *word processor* is no longer particularly accurate since most word processors include many desktop publishing features. Using word processors such as Microsoft Word 97, you can combine text, graphics, and equations within the same document.

Using the Mouse

You can use some of Word's features with either the keyboard or a mouse; however, some features are available only with a mouse. The *mouse* is a hand-held device that you use to make selections and issue commands. When you move the mouse, a *mouse pointer* moves correspondingly on the computer screen. As you move the mouse to the right, the pointer moves to the right; as you move the mouse to the left, the pointer moves to the left. The appearance of the mouse pointer changes depending on the operation you are performing and where you are working. While you are working in Word, the mouse pointer may take on any one of the appearances shown in Figure 2-1.

A mouse typically includes two or three buttons. In the Windows environment, you almost always use the left mouse button. All the mouse instructions in this text use the left mouse button unless otherwise noted.

Figure 2-1
Appearance of the mouse pointer

The first step in using the mouse is to *point* to the desired option or screen element, positioning the mouse pointer directly over it. Once you have done this, you either click, double-click, or drag, as described here.

- *Clicking* means pointing to the desired object and then pressing and quickly releasing the left mouse button.
- *Dragging* means pointing to the desired object, *holding down* the left mouse button while moving the mouse, and then releasing the mouse button.
- *Double-clicking* means pointing to the desired object and then clicking the left mouse button two times very rapidly. You must click fast enough and must not move the mouse between clicks for the selection to be made. If you have difficulty making selections by double-clicking, you are probably pressing down too hard on the mouse button or not clicking fast enough.

2-2 LEARNING YOUR WAY AROUND WORD

In this section, you learn how to start Word and acquaint yourself with the components of the screen. You also learn how to make menu selections and work in dialog boxes. Finally, you learn how to create, save, and print simple documents.

Starting Word

To start Word, look for the Microsoft Word *icon* like the one shown in Figure 2-2. An icon is a small graphic with text that is used in Windows-based applications to start the software or to change menus. Ask a lab assistant or your instructor for help if you cannot find this icon.

Figure 2-2
Microsoft Word icon

To start Word, double-click the icon.

In a moment, you will see a document window displayed, similar to the one shown in Figure 2-3. From here, you can create a new document or edit an existing one.

Figure 2-3
Word document window

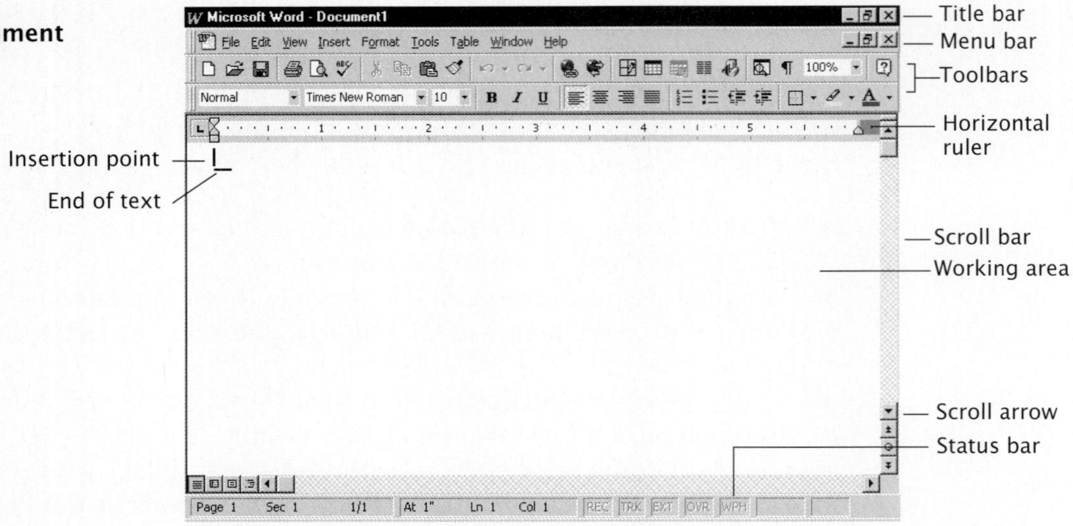

The Components of the Screen

To use Word effectively, you should get to know the components of the Word screen, most of which are labeled in Figure 2-3. Unlike DOS-based word processors you may have used, Word for Windows displays documents in *WYSIWYG* (pronounced *wiz-ee-wig*) mode. WYSIWYG stands for "what you see is what you get"—the way your document appears on the screen is very similar to the way it will look when printed. This feature enables you to see what your document will look like without obtaining unnecessary printouts.

The *insertion point* is the vertical line that identifies your current location within the document. As you type, the insertion point moves to the right and precedes your input. To move the insertion point to another location in your document, you can either use the arrow keys found on the keyboard or merely point with the mouse to the desired insertion point location and click. The small horizontal line shown just below the insertion point in Figure 2-3 shows the location of the bottom of the document.

As you are typing your document, you may wish to delete or erase part of what you have written. Using (BKSP) will delete characters to the left of the insertion point, whereas (DEL) will delete characters to the right of the insertion point.

The Title Bar At the top of the Word screen is the *title bar,* which names the application and provides information about the current document. When you first start Word, you will see a blank document, and the title bar will read "Microsoft Word - Document1," as shown in Figure 2-4. After you save your document and give it a name, the document name appears in the title bar.

Figure 2-4
Sample Word title bar

The Menu Bar Directly beneath the title bar is the *menu bar*, which is shown in Figure 2-5.

Figure 2-5
Word menu bar

If you click any of the menu bar options, a pull-down menu of additional options appears. These menus are one of your main means of issuing commands in Word. The types of options available in each menu are outlined here.

File	Manages your document (or files). From here, you can create a new document, save the document, print the current document, or exit Word.
Edit	Rearranges text and performs other editing tasks. Many of these editing options require that you first select a large block of text to be affected. You will learn how to select text in Chapter 3.
View	Customizes the way that your document looks on the screen as well as the appearance of the Word window itself. (The options do not change the way the document looks in print.) For example, you can magnify the document or display two pages side by side. You can also turn the ruler bar or the toolbars on or off.
Insert	Inserts special characters into your document. For example, you can insert the current date, footnotes, or captions. You can also bring text or graphics from other files into the current document.
Format	Controls the appearance of the document itself. (Remember, the View menu options change only the way the document looks on the screen.) The formatting options let you set the margins, text font, tab stops, text alignment, headers, number of columns, and more.
Tools	Includes the spelling checker, grammar checker, and online thesaurus. The Tools menu also features options for performing word counts or merging address lists with letters.
Table	Creates, edits, and customizes tables. You can add shading to table cells, split or join tables, and do more within a Word table.
Window	Switches between open documents or displays all open documents at once.
Help	Opens Word's online Help system. You can search for information about specific topics or tasks, get step-by-step instructions on how to use particular features, or access an extensive tutorial on the use of Word.

Toolbars Beneath the menu bar are the *toolbars*, which are shown in Figure 2-6. The top bar is the Standard toolbar; below it is the Formatting toolbar. Each button on the toolbars represents a Word command or function. You can access these functions by using the menu bar; however, if you are familiar with Word, it's usually faster to click the appropriate button than to go through the menu tree to make your selections.

Figure 2-6
Word default toolbars

Word has other toolbars available for your use, including Tables and Borders (contains buttons for including custom borders or shading in your document); Database (includes buttons that enable you to use Word as a simple database for tables of data); Drawing (includes buttons used to make simple line drawings in your document); Forms (contains buttons to create customized business forms and templates); Web (contains options for easy access to the Internet); and others. To change the toolbars displayed on your screen, choose Toolbars from the View menu, and then make the appropriate selection from the subsequent pull-down menu that appears.

You will learn about the buttons on the Standard and Formatting toolbars throughout this module, as the commands or functions they represent are introduced. In most cases, a button's function is clear from its icon. If it is not, you can simply point to the icon, and Word will display a brief description of the icon's function.

The Horizontal Ruler Directly beneath the toolbars is the *horizontal ruler*, as shown in Figure 2-7. The horizontal ruler is used to set tab stops and right and left page margins.

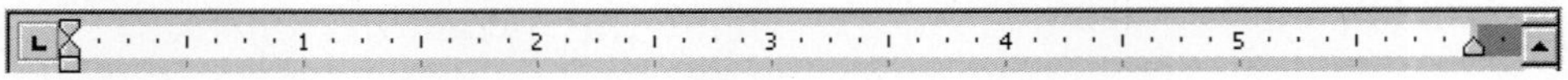

Figure 2-7
Word ruler

The Text Area The main part of the Word screen is called the *text area*. It is here that you create and view your document. The page margins are automatically set for you within Word. By default, the top and bottom margins for a document are set to 1 inch, and the right and left margins are set to 1.25 inches. The next chapter explains how to change the margins.

The Status Bar At the bottom of the Word document window is the *status bar*, which is shown in Figure 2-8. The status bar displays general information about the document you are working on. The left side of the status bar provides information on the text that is currently visible within the display window. When you start Word, this will be displayed as Page 1, Sec 1, and 1/1 (Page 1 of 1). As you create your document, these numbers will change, depending on the text that is currently visible.

The middle section of the status bar shows the current location of the insertion point. The first number indicates the location of the insertion point relative to the top of the paper. Initially, this will be At 1″ (recall that Word automatically set a top margin of 1 inch). The second number indicates the line number of the insertion point relative to the top of the document. The last number lists the column position within the line of text.

On the right side of the status bar, information about various typing modes is shown. The five available mode indicators are used to record macros, to select text in extend mode, to type in insert or overtype mode, and to use revision marks. The last mode indicator can be used to obtain help for WordPerfect users who are trying to learn Word. A final indicator is available to easily check your spelling and grammar checking status.

Figure 2-8
Word status bar

The Scroll Bars Like most Windows applications, Word includes *scroll bars* along the right side and the bottom of the document window, as shown in Figure 2-9. You can use these scroll bars to move through your document. To move up, click the up arrow on the vertical scroll bar; to move down, click the down arrow. You can also use the horizontal scroll bar to move to the left and right across the page.

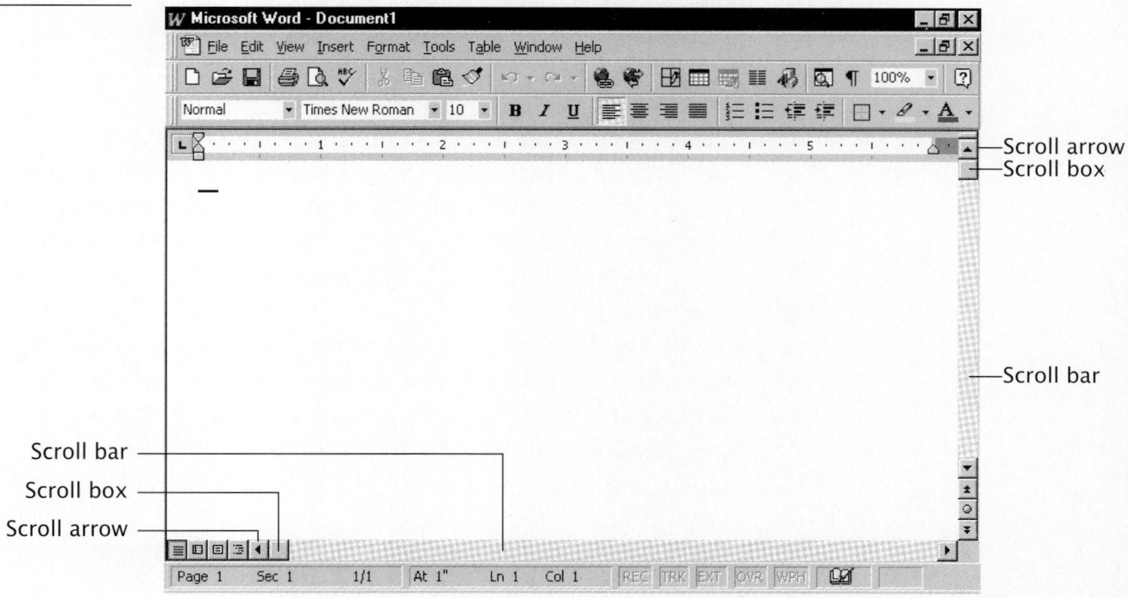

Figure 2-9
Scroll bars

Both scroll bars contain a *scroll box* that indicates your position in the document. For example, if the scroll box is at the very top of the scroll bar, you will see the top of the document. You can drag the scroll box within the scroll bar to move quickly to another location in your document.

At the left on the horizontal scroll bar are four buttons, shown in Figure 2-10. These buttons control how the document is viewed on the screen. The Normal view button is used for most typing and editing. This viewing option is set by default for you in Word. The Outline view button is used to show only the main headings of the document so that you can quickly scroll through it. The Online Layout view button is used if you are reading your document online. Larger fonts as well as increased space between lines are used for increased legibility. The Page Layout view button is used to preview your document so that you can make sure it appears as you want before printing.

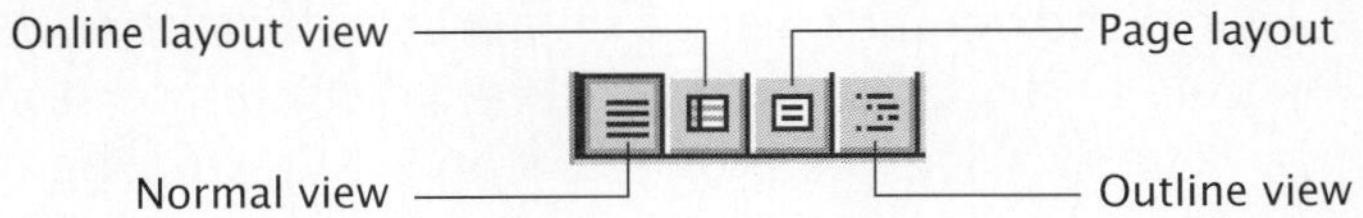

Figure 2-10
Page view buttons

Using Menus and Dialog Boxes

You display pull-down menus by clicking the desired option in the menu bar. Once a pull-down menu appears, you can select a menu option by clicking it. After making your menu selection, dialog boxes will appear on the screen. *Dialog boxes* communicate with Word, providing the program with additional information that it needs to carry out a command. Most of the time, the options you specify within a dialog box are self-explanatory. Many dialog boxes include some default settings, which you are free to change before executing the command.

Typically, after you finish making changes in the dialog box, you click OK to execute the commands and close the dialog box. You can also click Cancel to close the dialog box and return to your document without issuing the command. Throughout this text, you will learn about various dialog boxes as you create and edit documents.

Using Templates

Many times when you are working in Word, you will need to create a new document without exiting Word or closing your existing document. If you choose New from the File menu, a dialog box like the one shown in Figure 2-11 appears. The contents of this dialog box include several templates available in Word. A *template* is a document that contains text and paragraph formats that are already set up for you. If you use these templates (some are called Wizards), you will be able to create professional-looking documentation with less effort. Some of the document templates include standardized forms for letters, memos, resumes, reports, and others. A complete description of available templates in Word is beyond the scope of this text. If you would like to use any of these templates, consult either

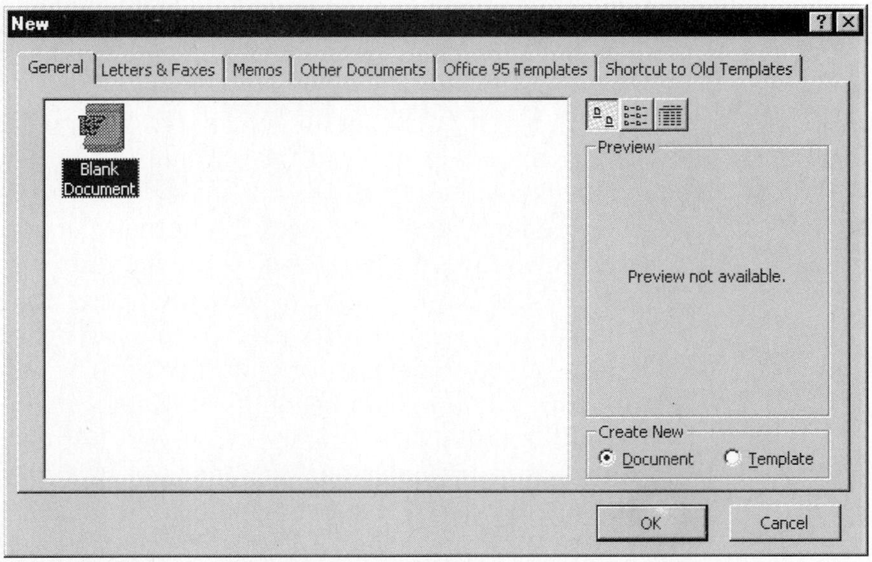

Figure 2-11
New dialog box

the online Help or the Word manual for more information. The default template that is displayed when you start Word is the General Blank Document template.

Saving a Document

If you wish to keep a document you have created, you must save it by choosing either Save or Save As from the File menu. Use Save for documents that you have previously created; Save overwrites the former version of the document with the updated version. To save a document for the first time or to save a copy of your current document under a different filename, select Save As. You will see the Save As dialog box similar to the one shown in Figure 2-12. You will enter document names in the File Name text box and then identify the proper directory in which to store the file. You can also save your document as a different file type by selecting the appropriate option from the Save as type drop-down list box. Some of the available file types include text only and MS-DOS text.

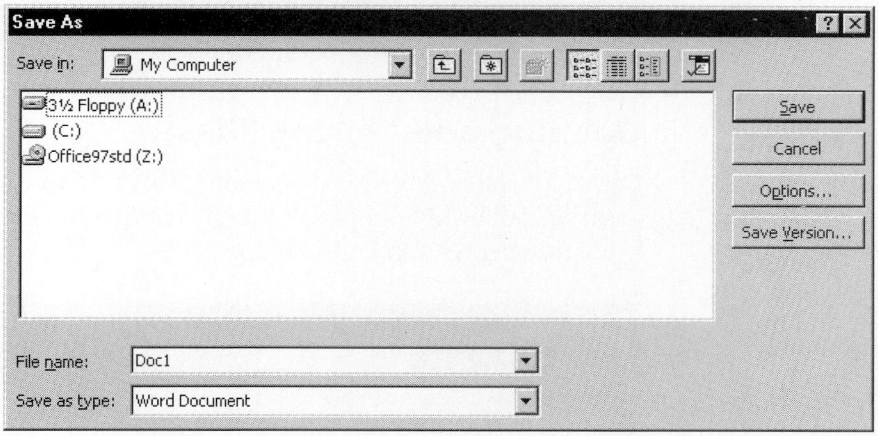

Figure 2-12
Save As dialog box

Note: Choosing Save or Save As displays a dialog box with a default filename for the current file. Be careful not to use the name of an existing file, or you may overwrite the file of that name.

When you name your documents, you will usually include a filename followed by a period (.) and then an extension. Typically, the extension signifies the file type, and the name signifies the file contents. The most commonly used file extension for Word documents is `.DOC`. You should establish your own consistent filenaming conventions. Having such a system makes it relatively easy to retrieve documents that you created some time ago.

After you name your file, you need to tell Word where to save it. You can specify the disk drive and the directory within the dialog box. (In Figure 2-12, drive A: and C: are available.) To select a drive, double-click the desired drive from the list.

To choose a new file folder, double-click the appropriate folder in the display box. Note that double-clicking a folder displays all the folders and files it contains. You may need to do this if the folder you want is not initially displayed in the box.

After you name your document and choose where to save it, select OK to save the document to the appropriate location on the disk.

Try It In this exercise, you create a short document and then save it.

- Start Word using the procedure described in this chapter or the method used in your particular lab.
- Type the following sentences:

  `In the engineering profession, the synthesis of knowledge is of paramount importance. Engineers are involved in the design, analysis, and manufacture of new or improved products. Engineers employ mathematical and scientific discoveries to construct the products that improve our everyday lives.`

  As you type, note that the text automatically "wraps" to the next line—you do not need to press ENTER at the end of each line.

- When you are finished typing these sentences, press ENTER two times, and then type your name and the date.
- Save the file as TRIAL1.DOC on drive A: or B:.

Opening and Closing Files

Once you have saved a document, you can open and edit it. To open an existing document, choose Open from the File menu. The Open File dialog box appears, as shown in Figure 2-13.

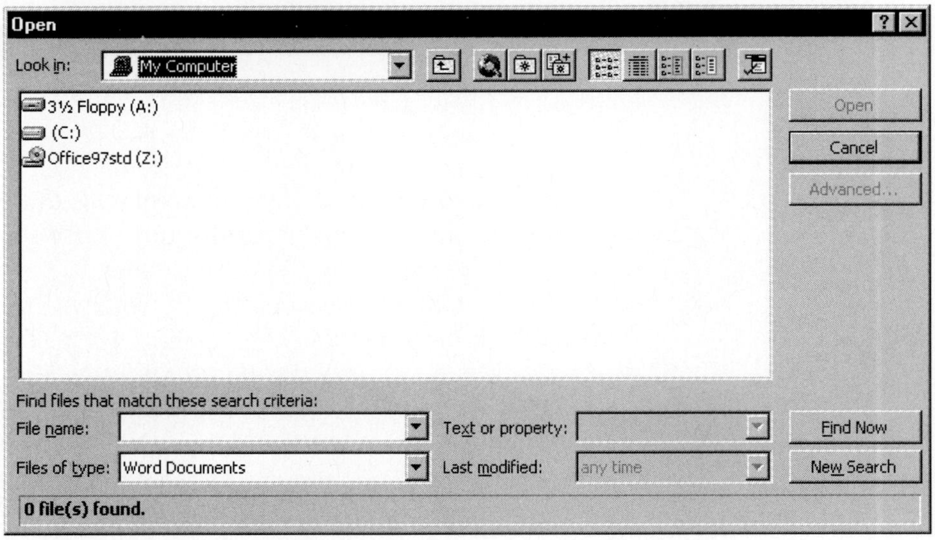

Figure 2-13
Open dialog box

To open a file, first specify the drive and the directory where the file resides. You choose a new folder from the Look in drop-down list box; this is just like selecting a new folder in the Save As dialog box. To change folders, double-click the desired folder. Word displays the contents of the selected folder in the folder/file display box. To open a particular file, double-click its folder.

Often you will need to remove the current document from the screen, but you must remain in Word to work with other documents. To close the current document and remain in Word, choose Close from the File menu. If you have changed the file since you last saved it, Word asks whether you wish to save the changes to the file before closing it.

Try It In this exercise, you practice opening and closing files.

- Close the current document, and start a new document.
- Type the following sentences:

  `To work successfully, engineers must be able to effectively communicate their ideas in both written and graphical form. Throughout the design and implementation process, engineers express their ideas with both drawings and written documents.`

- Save this file with the filename `TRIAL2.DOC`, and then close it.
- Reopen the file `TRIAL1.DOC`.

Printing a Document

After creating a document, you may want to print it. Figure 2-14 shows the dialog box that appears when you choose Print from the File menu.

Notice the default options: Document is the default in the Print What list box, All is the default for Page Range, and so on. The default printer name also is listed in the dialog box. (Your default printer will most likely differ from the one listed in Figure 2-14.)

Before you print your document, make any needed changes in the Print dialog box. For example, click the down arrow near the Printer list box if the printer you wish to send your document to is not the same as the default. A listing of all the available printers that Word supports in your laboratory setting appears. You can choose one of these printers by clicking its name. You can also change the number of copies or the type of document you wish to print, the range of pages, or any of the other options listed on the screen.

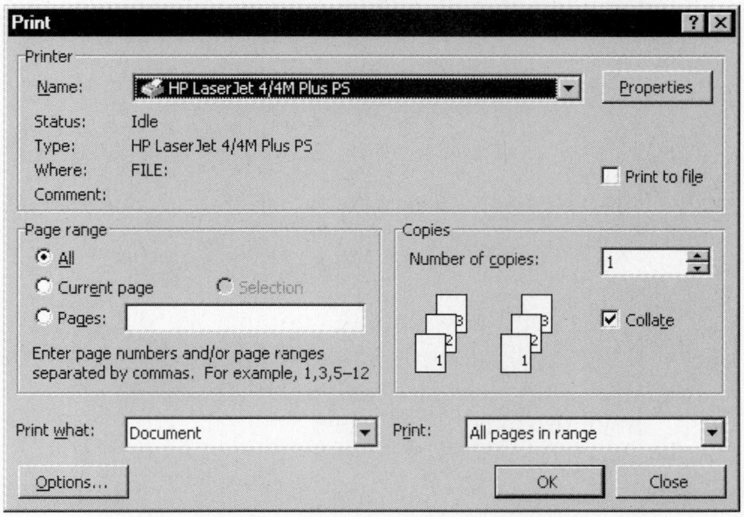

Figure 2-14
Print dialog box

When you are finished making selections in the Print dialog box, choose OK to send your document to the printer. If you decide not to print any thing at this time, click the Cancel/Close button.

You may experience problems printing your documents, depending on the printer you're using. Some printers, for example, may not be able to print some of the fonts that you use in your document. If you have trouble printing a document, contact a lab assistant or your instructor.

Try It

In this exercise, you print the document you created in the first Try It! exercise.

- Open TRIAL1.DOC if you have not already done so.
- Print the document, making sure that the correct printer is specified. (You may want to check with your instructor.)

Getting Online Help

When you need help, you can refer to Word's *online Help* feature. Online Help explains how to perform various Word operations. To open the Help window, choose Help from the menu bar. The Help pull-down menu shown in Figure 2-15 appears.

Figure 2-15
Getting online Help in Word

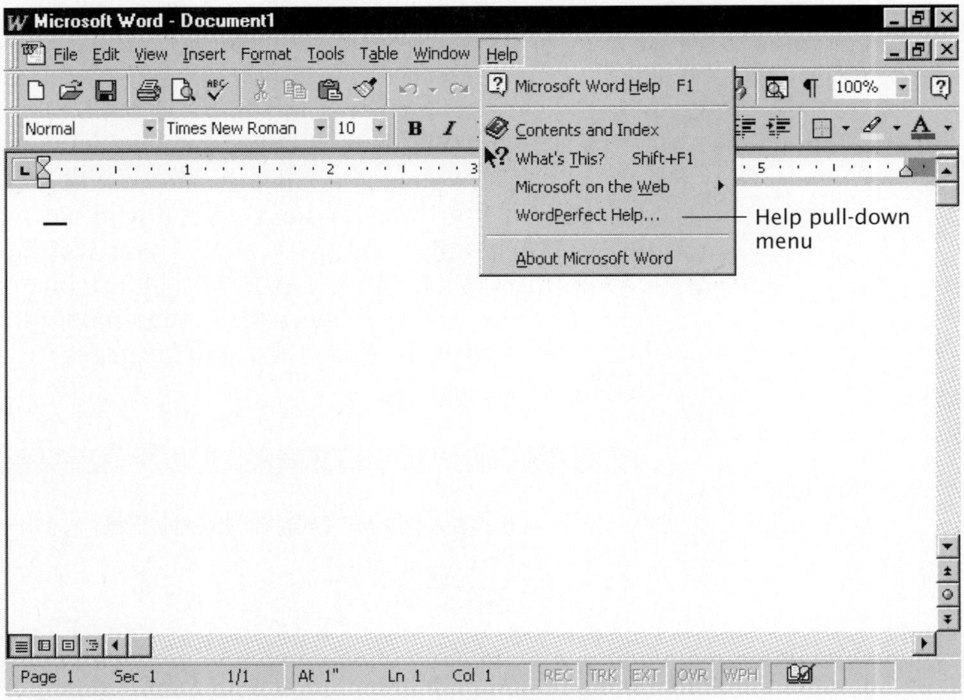

This pull-down menu contains Word's various Help options. One of the most useful features here is the Contents and Index option. If you select this option, the Help Topics: Microsoft Word dialog box appears, as shown in Figure 2-16.

Figure 2-16
Help Topics: Microsoft Word Dialog Box

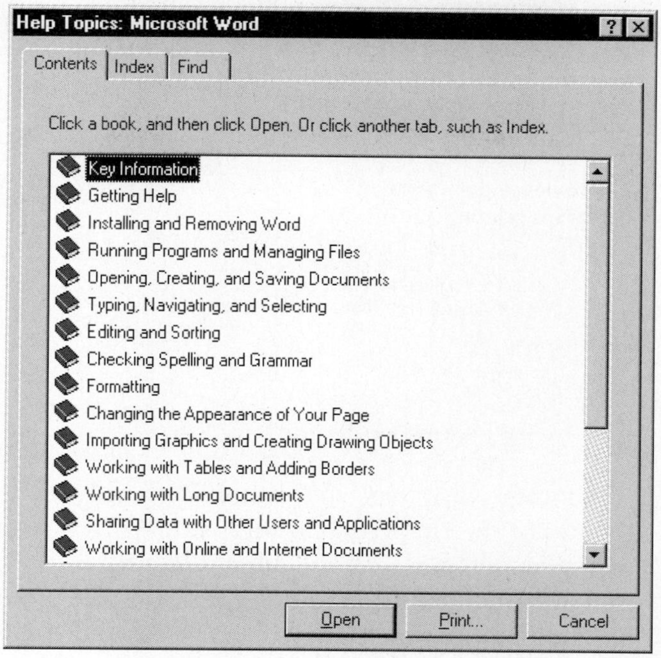

When you double-click a topic, a listing of all the relevant help topics is displayed. You then double-click the topic from this new list to display further help topics. For example, double-clicking the topic Printing displays a listing of three available options for help in printing—Previewing a Document Before Printing, Printing Your Work, and Creating and Printing Envelopes and Labels. If you then double-click the topic Previewing a Document Before Printing, the sub-list of this option is displayed as shown in Figure 2-17.

Figure 2-17
Previewing a Document Before Printing Help List

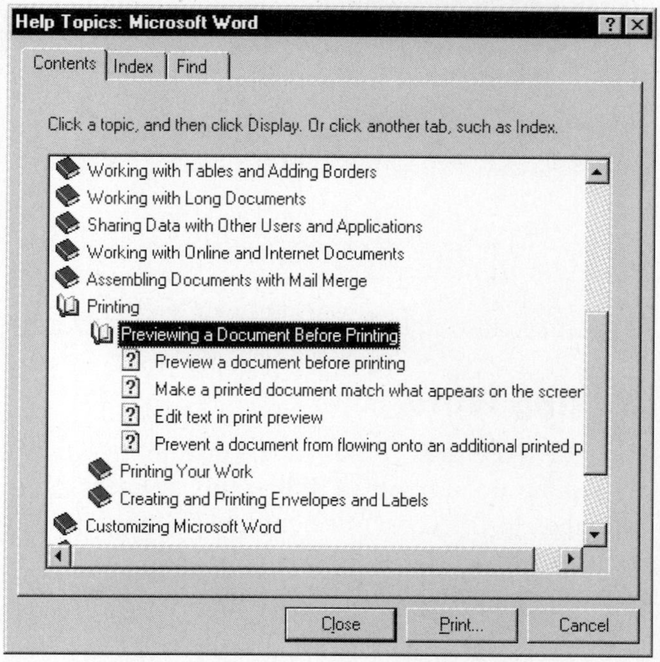

If you now select the topic Preview a document before printing, the Help window shown in Figure 2-18 appears.

Figure 2-18
Preview a document before printing Help window

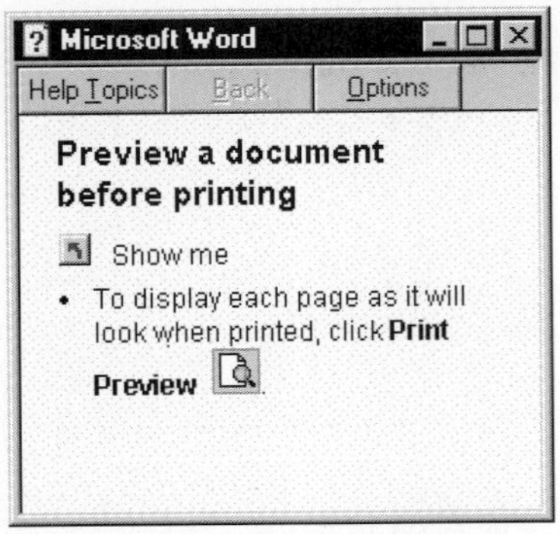

Another way to use the online Help in Word is to click the Index tab at the top of the Help Topics: Microsoft Word dialog box. The contents of the dialog box change and will appear like that shown in Figure 2-19. Here you can type the first few characters in a topic for Help. The list in the dialog box will change accordingly. When you see the desired Help topic within the list box, double-click the topic, and Help on that topic is displayed for you.

Figure 2-19
Help Topics: Microsoft Word Index option

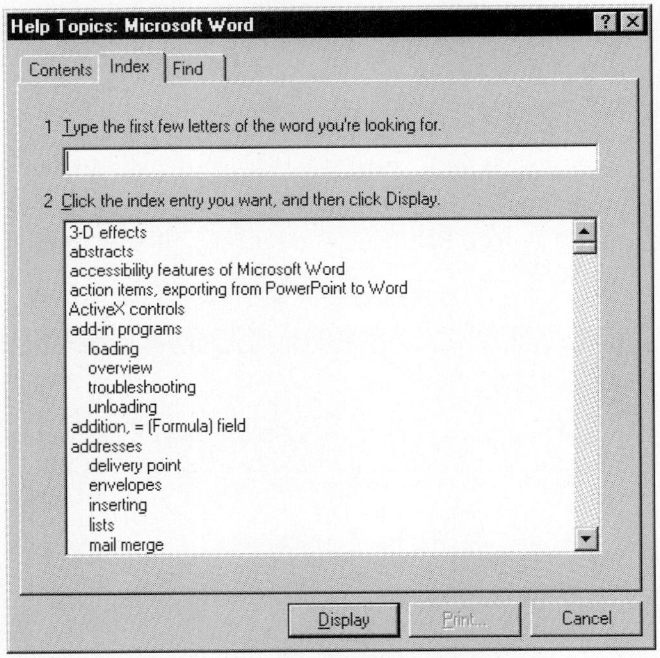

Exiting Word

To exit Word, select Exit from the File menu. If you have made changes to any open documents, a dialog box appears asking if you want to save the changes to the current document. Choose Yes to save your changes, No to discard the changes, Cancel if you decide not to exit Word, or Help if you need help saving your document.

Try It In this exercise, you open the online Help system, learn how to check a document's spelling and grammar, and then exit Word.

- Choose Contents and Index from the Help menu.
- Select the topic Checking Spelling and Grammar, and then select the Correct Spelling and Grammar option.
- Close the Help windows and exit Word.

2-3 USING SHORTCUTS

As noted in Section 2-2, you can issue many Word commands more easily by clicking the appropriate buttons on the toolbars. The following buttons on the Standard toolbar can be used to access four commands introduced in this chapter:

 Starts a New document

 Opens a document

 Saves a document

 Prints a document

SUMMARY

In this chapter, you learned the basics of how to start Word 97 and how to navigate around the Word screen. Then you learned how to create and save a new document, open and close existing documents, and print documents. Finally, you learned how to get online Help in Word.

Key Words

click
dialog box
document window
double-click
drag
horizontal ruler
icon
insertion point
menu bar
mouse
mouse pointer
online Help
point
scroll bars
scroll box
status bar
templates
text area
title bar
toolbars
word processing
WYSIWYG

Exercises

1. Start Word. Type five sentences about the engineering field you will major in and why you prefer this field. Print the document.
2. Open the file you created as a Try It! exercise, `TRIAL2.DOC`, and modify the second sentence to read

   ```
   Throughout the design and manufacturing process,
   engineers communicate their ideas through drawings and
   written documents.
   ```

   Obtain a printout of the new document.
3. Write a paragraph describing your long-term career goals. Print and save this file.
4. Read some of the Help information regarding Formatting Text Characters, and summarize what it says in a few sentences within a new Word document. Print this file.

3

Writing Letters and Memos

Space Travel

On January 20, 1969, the world looked on in awe as we made one "giant leap for mankind." As Neil Armstrong and Buzz Aldrin made their historic first walk on the moon, families across the nation and across the world watched and listened. Communications was a significant problem for the space exploration program. There had to be communication links between the men on the moon and the Apollo spacecraft, and between the spacecraft and mission control on Earth. These communication links were vital to solving any problems that developed during the flight. Although the astronauts could solve many electrical or mechanical problems as they occurred, they did not have the expertise to handle all the potential problems associated with such a complicated mission. The communication link with home base was necessary to guarantee the success of the mission. In this chapter, we'll explore the two types of documents engineers use to communicate during the first stage of the engineering design process: defining the problem.

INTRODUCTION

This chapter focuses on the first stage in the engineering design process—identifying the problem—and discusses the two types of documents engineers typically create during this stage: letters and memos. You learn how to create simple letters and memos in Word and how to customize the appearance of these documents.

3-1 DEFINING THE PROBLEM

Perhaps the most critical stage of the engineering design process is the first one: defining the problem completely and succinctly. Problem definition is especially important in engineering because there are innumerable solutions to any one problem, and the engineer must choose the best solution from the almost limitless possibilities. If the engineer does not thoroughly understand the problem, the proposed solution is not likely to be the optimum solution.

For example, suppose you are an automotive engineer who receives word from one of your service centers that a certain gear within an engine is failing frequently. Before you can solve this problem, you must understand what is causing the part to fail. Here are some possibilities:

- The original design was faulty, and the gear cannot withstand the applied loads.
- Incorrect material was used in its manufacture.
- The material supplied to the manufacturer did not meet the specifications of the design engineer.
- The manufacturing processes introduced microscopic flaws in the part.
- Other engine parts that interact with this gear were incorrectly designed and are overstressing the part.
- Consumers are using the vehicle in a way that was unforeseen in the original design process.

Clearly, you need to determine which of these possibilities is the real culprit before you can solve the overall problem—that the gear is subject to frequent failure. To do so, you need to communicate effectively with everyone involved in the design, manufacture, and use of this gear. In this stage of the design process, the most common form of documentation is communication in the form of letters or memos. The remainder of this chapter describes how to use Word to write letters and memos.

3-2 COMMUNICATING WITH LETTERS AND MEMOS

Letters and memos are used for direct communication with others. Unlike personal letters, business letters and memos are typically direct and to the point. Frequently, such letters and memos are single-purpose documents written to create a "paper trail" throughout the design process. For example, if you are an engineer in a consulting firm and are responsible for the design of a city parking deck, you would probably start out by asking your client to express, in a written letter or memo, the required capacity for the parking deck, its location, and other characteristics. In this way, everyone has a clear idea of the project requirements, and, if questions arise at a later date, there is a written record of the stated initial requirements.

One of the primary differences between letters and memos is their intended audience. In general, memos are written to people working within your office or company, whereas letters are written to people employed outside your office. Letters and memos also have different formats.

Letter Format

In the business world, letters are usually written on company stationery (letterhead) that includes the company name, address, and possibly the phone and fax numbers. When using Word to write letters, you can either print or photocopy the text of the letter onto the company letterhead. Figure 3-1 shows a typical business letter.

Figure 3-1
Sample business letter

Letterhead — XYZ Consulting 1055 Washington Boulevard Fort Wayne, IN 51001

Date — January 2, 1998

Name and address of recipient — Ms. Helen Johnston
Staff Engineer
United Pump Technologies
120 Avenue G
Evanston, Illinois 53321

Dear Ms. Johnston,

Body — As you requested by phone yesterday, I am sending you somple information regarding the reliability of the United Pump Technologies Model SR151. Our company has used this particular pump model in numerous well and water system designs in the Fort Wayne area. We have had good luck with the pumps in most installations. However, in certain locales they have been known to "freeze up" and had to be replaced. Most of the failed installations have been in sandy soil types, although this is not true of all of them. I am also shipping you one of the pumps that failed.

Please let me know if I can be of further assistance.

Closing — Sincerely,

Bob Griffin

Bob Griffin
Project Engineer

Because the letterhead already contains the name and address of your company, you do not have to include this information in the letter. However, for documentation purposes, you should always include the date on your letters.

Include the name and address of the letter recipient below the date. If you don't know the person's name, you can address the recipient as "Dear Sir/Madam" or "To Whom It May Concern."

The information in the body of the letter should be direct and to the point. When writing to someone you have never met or spoken with, it is customary to introduce yourself at the beginning of the letter. You may begin the letter by saying something like "I am a senior project engineer with the Braun Corporation in charge of the construction of the new city sewage facility." If the letter is a written response to an earlier verbal communication, you might start with something like "As you may recall from our phone conversation of January 15 ..." You should always close the letter by signing your name and clearly typing your name underneath in case your signature is illegible.

It is important that you check your letter for spelling and grammatical errors. (Word's spelling and grammar checking tools are discussed in Section 3-4.) And since Word's spelling and grammar checkers are by no means infallible, you should proofread the letter before mailing it.

Memo Format

Memos are not as formal as letters, but they are equally important in the documentation of an engineering design. A company memo form (similar to the formal company stationery) is not required, but it may be used.

Figure 3-2 shows a typical memo format. Because memos are written to individuals within your own company or office, the address of the recipient is not needed, and the writing style is not usually as formal as the style in a letter. After RE: (short for "in regard to"), insert a brief description of the purpose of the memo. In memos, as in letters, it is important to include the date for documentation purposes.

Figure 3-2
Sample memo format

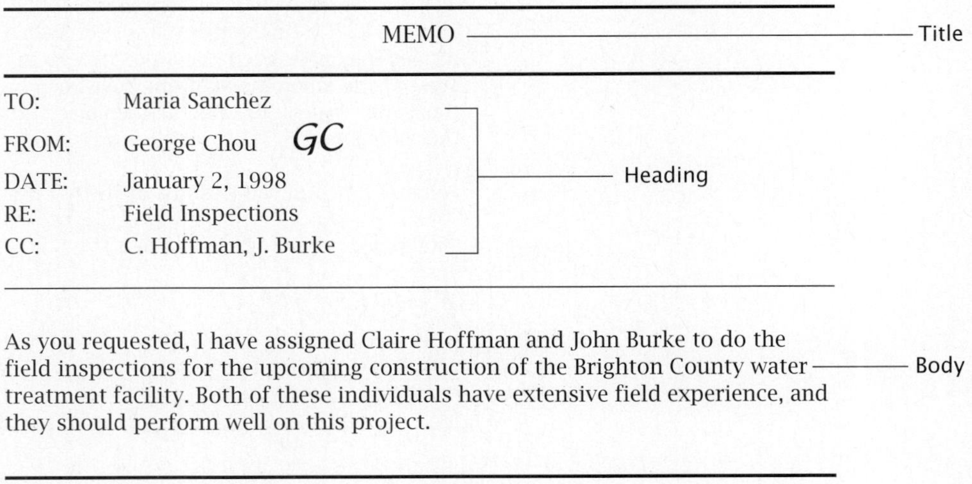
MEMO

TO: Maria Sanchez
FROM: George Chou GC
DATE: January 2, 1998
RE: Field Inspections
CC: C. Hoffman, J. Burke

As you requested, I have assigned Claire Hoffman and John Burke to do the field inspections for the upcoming construction of the Brighton County water treatment facility. Both of these individuals have extensive field experience, and they should perform well on this project.

Sometimes you need to send a memo to more than one person. If so, include the names of everyone who should receive a copy of the memo on the line labeled CC:. This is an abbreviation for "carbon copy" and reflects an earlier era when photocopiers were not available and carbons of documents were the norm. It is not usually important to sign memos, although it is good practice to initial the memo next to your name.

3-3 CREATING LETTERS AND MEMOS IN WORD

This section examines Word options that you can use to create letters and memos. You will learn many ways of modifying the appearance of these documents, including how to modify margins, tabs, indents, and line spacing. You will then learn how to change font characteristics such as the font face and font size. Finally, you will learn how to use Word to check your document for spelling and grammatical errors.

Setting Margins

Word has default margins of 1.25 inches on each side and 1 inch on the top and the bottom. You cannot type or place the insertion point anywhere in the margins. If you wish to change a document's margins, choose Page Setup from the File menu. The Page Setup dialog box appears, as shown in Figure 3-3. Note that there is a separate text box for each margin—left, right, top, and bottom.

Figure 3-3
Page Setup dialog box

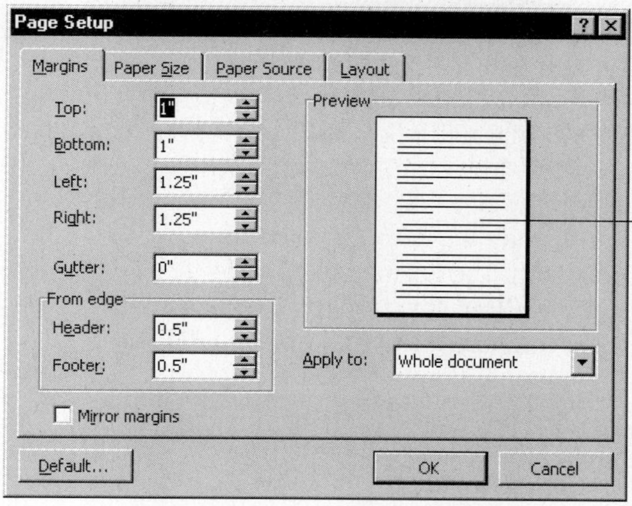

Preview page display

To change any or all of the document margins, click within the desired text boxes, delete the current margin setting, and type in a new one. You can also point to either arrow next to the desired margin; click the up arrow to increase the margin size, or click the down arrow to decrease the margin size. This method changes the margin in increments of 0.1 inch. As you change the margins, the Preview page display changes to reflect the new settings. For example, Figure 3-4 shows the Page Setup dialog box after the left margin setting has been changed to 2 inches and the top margin setting has been changed to 3.5 inches. To close the dialog box and apply these margins to your working document, choose OK.

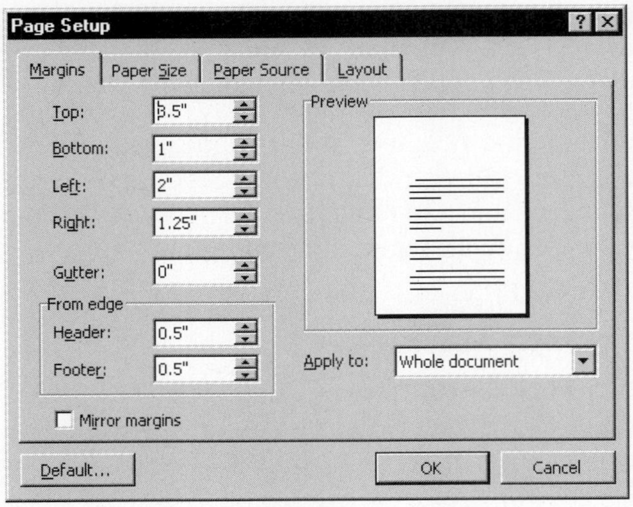

Figure 3-4
Page Setup dialog box after specified changes

Setting Tab Stops

The *tab stops* in your document cause the insertion point to "jump" to a predefined location within the margins when you press TAB. For example, suppose you have a tab stop set at 4.5 inches in from the left margin. If you are at the beginning of a new line and you press TAB, the insertion point will automatically move to the point in that line that is 4.5 inches from the left margin. By default, Word documents have an initial tab stop set at 1/8 inch from the left margin, with additional tab stops every 1/2 inch across the page. By default, the tab stops are left-aligned.

You may want to change these default tab settings. To do so, choose Tabs from the Format menu. The Tabs dialog box illustrated in Figure 3-5 appears. You can use the options in this dialog box to create new tab stops or change existing ones.

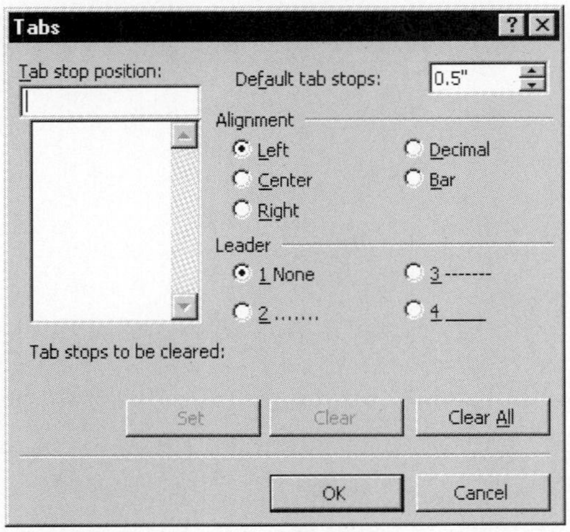

Figure 3-5
Tabs dialog box

To change the position of tab stops, type a new value for the tab settings in the Tab Stop Position text box and then choose Set. To change the alignment of a tab stop, select the tab stop you wish to change, and then select the Alignment option in the dialog box. As noted, the default is left-aligned. Other options are right-aligned, centered (text is centered around the tab stop), and decimal (numbers are aligned on the decimal point).

You can also use the Bar option found in this dialog box to create a vertical line at the location from the left margin that you specify. The Leader option allows you to fill the empty space preceding a tab stop with a dotted, dashed, or solid line.

After you have created tab stops, the horizontal ruler's appearance will change to show you the type and location of each tab. Figure 3-6 shows the ruler with a left tab set at .75 inch, a centered tab set at 2.0 inches, a right tab set at 3.25 inches, a decimal tab set at 4.25 inches, and a bar set at 5.0 inches. After these tab stops are visible in the ruler, you can change their location by dragging any of the tab symbols to a new location along the ruler.

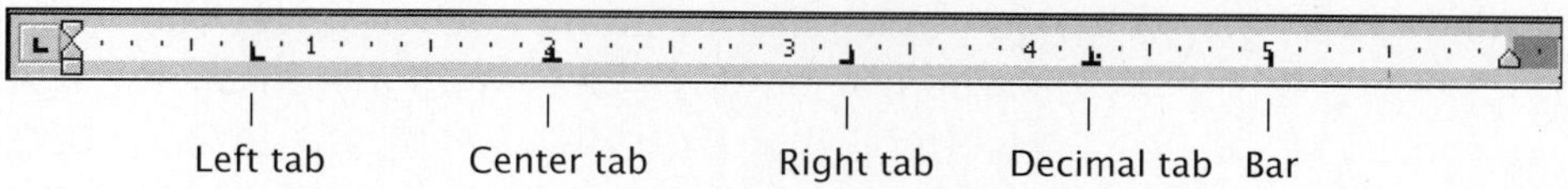

Figure 3-6
The Word ruler with tab stops included

Inserting the Date in a Document

Remember that you should include the date in all engineering documents. This is easy to do with Word, assuming that your computer's clock is set correctly. To insert the date in your document, position the insertion point where you want to place the date, and choose Date and Time from the Insert menu. A dialog box similar to the one shown in Figure 3-7 appears. You insert the current date and/or time by selecting the display style from the list shown in the dialog box and then choosing OK.

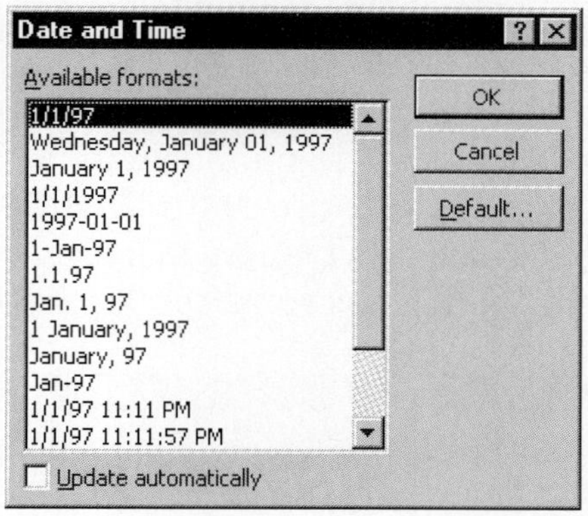

Figure 3-7
Date and Time dialog box

Try It

In this exercise, you create a memo heading similar to this.

MEMO

TO: Instructor's Name

FROM: Your Name

DATE: Current Date

RE: Problem Solving

- Start Word if necessary.
- Change the top margin to 2 inches and the right margin to 1.5 inches.
- Clear all tabs.
- Set a tab stop at 0.75 inch from the left margin.

- Type MEMO, and press ENTER two times.
- Type TO:, press TAB, type the name of your instructor, and press ENTER two times.
- Continue entering text to create the required heading. To include the date in your heading, use Word's Insert-Date feature.
- Save the document as `MEMO1.DOC`.

Selecting Text

Before you learn more about how to customize your document's appearance, you need to know how to select the text that you want to change. *Selecting text* simply means highlighting the text you want to modify. To select text in a Word document, just drag across it. (Remember, dragging means positioning the mouse pointer at the place where you want to make changes, pressing and holding down the mouse button, moving the mouse, and then releasing the mouse button when the mouse pointer is in the next desired location.)

Specifying Paragraph Formats

A paragraph is a block of text that is set off by a hard carriage return from the text above and below it. A hard carriage return is inserted into your document each time you press the ENTER key on the keyboard. You can use the options in Word to customize the appearance of individual paragraphs or blocks of paragraphs. To simplify editing your document, it is important that you not insert any unnecessary hard carriage returns.

There are several Word options for customizing the appearance of paragraphs. You can set the paragraph indentation, the line spacing, and the justification. All these features are controlled by options found in the Paragraph dialog box. To change a paragraph's format, choose Paragraph from the Format menu. A dialog box similar to Figure 3-8 appears.

Figure 3-8
Paragraph format dialog box

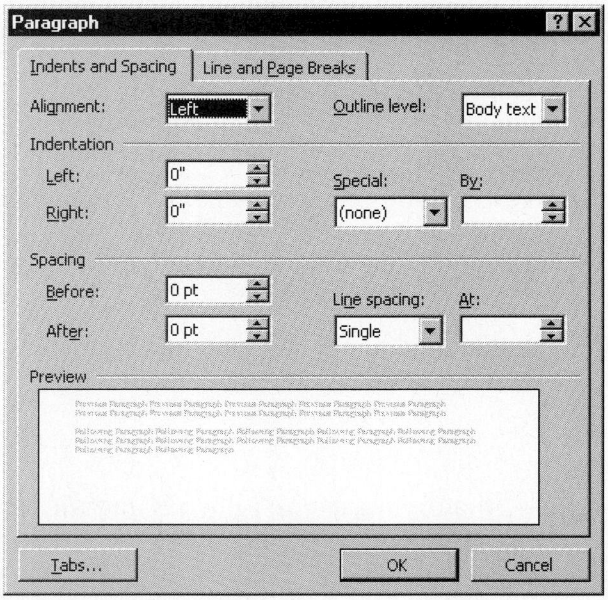

Indenting Paragraphs You can use four types of *indentation* to customize the look of your document. Left indentation offsets all lines of a paragraph a preset distance in from the left page margin; right indentation, from the right margin. If you choose the Special drop-down list box, you can choose from two more types of indentation: first line and hanging. With first line indentation, the first line of a paragraph is indented, and all other lines in the paragraph are aligned along the left margin. Hanging indentation means that the first line of the paragraph is aligned along the left margin, and all the remaining lines in the paragraph are indented from the left margin. Whatever type of indentation you choose, you specify the distance from the margin in the appropriate text box. Figure 3-9 illustrates the four types of indentation.

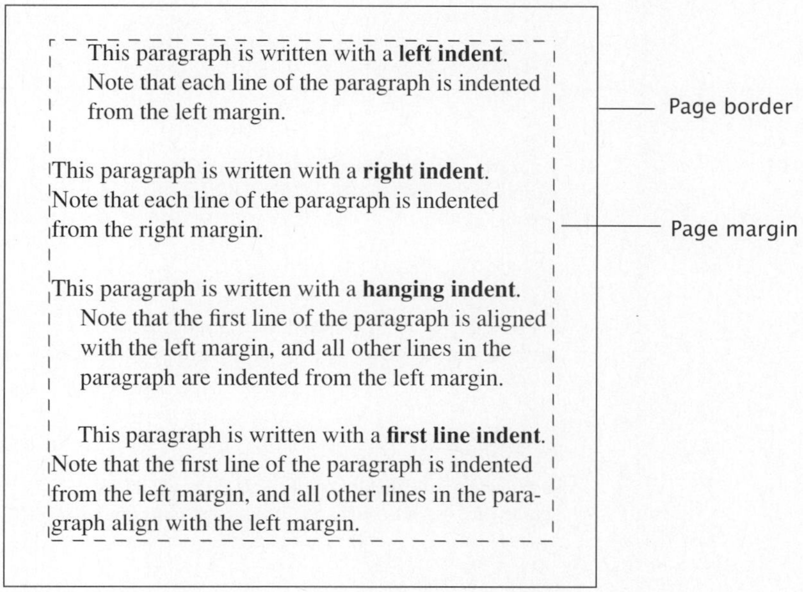

Figure 3-9
Word's indent features

Specifying Line and Paragraph Spacing The *line spacing* of a paragraph governs the vertical distance between lines of text. The *paragraph spacing* governs the vertical distance between paragraphs in the document.

Single spacing is commonly used as the line spacing for letters and memos; however, most other types of engineering documentation use double spacing. The default line spacing in Word is single spacing. The line spacing for a paragraph is also set in the Paragraph dialog box (see Figure 3-8). To change the line spacing, choose one of the options from the Line Spacing drop-down list box. The three most common types of line spacing are Single, 1.5, and Double. Other options are also available for customizing your line spacing.

Paragraph spacing can be changed to include extra space before and/or after each paragraph. The default paragraph spacing is single spaced both above and below the paragraph. Paragraph spacing is defined by changing the setting in the Before or After text boxes. The spacing is defined in terms of point sizes. There are 72 points in an inch, and standard text size is 10 or 12 points. Thus, if you set a spacing of either 10 or 12 points *before* a paragraph, this will have the same effect as inserting a blank line between this paragraph and the preceding one.

Specifying Paragraph Justification *Justification* refers to the alignment of text relative to the right and left margins. Word provides four types of paragraph justification. When text is left-justified, it aligns with the left margin and is “ragged” (uneven) along the right margin. When text is right-justified, it aligns with the right margin and is ragged along the left margin. With center justification, each line of text is centered between the right and left margins. Justified means that all lines of text in the paragraph are aligned along both the right and left margins (space is added between the words if necessary). The Word default is left justification.

To change the justification of the current or selected paragraphs, choose from the Alignment drop-down list box found on the Paragraph dialog box (see Figure 3-8). Figure 3-10 illustrates the different types of justification available in Word.

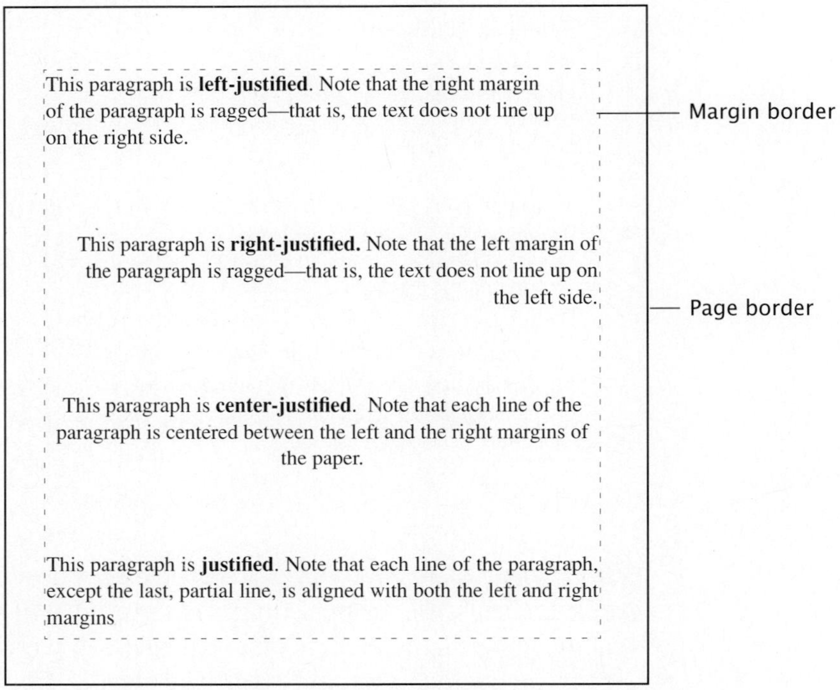

Figure 3-10
Paragraph justification options

Changes in indentation, justification, and line and paragraph spacing affect the current or selected paragraphs only, and not the entire document.

Try It

In this exercise, you continue to work with the memo you started in the previous Try It! exercise.

- Open MEMO1.DOC if it is not already open. Type the following short paragraphs as the body of your memo. (Make sure you include a hard carriage return at the end of each paragraph.)

```
Engineers solve problems virtually every day of their
careers. Some of the problems encountered on the job are
relatively trivial, whereas others are extremely complex.
You can solve most problems by using the five-step prob-
lem-solving method outlined here.

1. Define the problem. It is extremely important that you
   understand the problem before attempting to determine
   an optimal solution.
2. Gather the data. When solving a problem, you frequently
   start with incomplete data. It is critical to amass all
   the necessary data before beginning to come up with a
   solution.
3. Look at alternative solutions. Most engineering prob-
   lems have several possible solutions. You need to con-
   sider all the alternatives before committing to a final
   solution.
4. Select the "best" solution. After carefully weighing
   all the possible solutions, select the solution that
   satisfies all or most of the design parameters.
5. Implement the solution. Virtually all engineered solu-
   tions are either constructed or manufactured as the
   final stage in the problem-solving process.
```

- Place the insertion point in the paragraph that consists of the word MEMO. Change the justification to center this paragraph.
- Select the five paragraphs that outline the problem-solving steps. Change the indentation of these paragraphs to hanging indent. Specify an indentation of 0.2 inch from the left margin.
- Save the memo under the filename MEMO2.DOC.

Specifying Font Formats

In Word, text can be customized by any of four different characteristics—size, font, style, and special effects. If any text is selected when you change a characteristic, only that text is affected. If no text is selected, the characters that you subsequently type will be affected by your changes.

Specifying the Font and Size The *font* is the lettering style. Your printer may not support all the Word fonts. The font size is specified in units called *points*. Recall that there are 72 points to an inch (a 72-point font is 1 inch high). The default font size in Word is 10 points. Text sizes ranging from 10 to 12 points are considered standard for creating most types of documentation. Figure 3-11 illustrates some of the more commonly used fonts and sizes.

Figure 3-11
Commonly used font faces and font sizes

This sentence is written in 12-point Times New Roman.

This sentence is written in 10-point Helvetica Narrow.

This sentence is written in 14-point Bookman.

This sentence is written in 12-point Courier.

This sentence is written in 18-point Helvetica.

This sentence is written in 14-point Avant Garde.

This sentence is written in 24-point Palatino.

Specifying the Font Style and Appearance The *font style* includes options such as boldface, italics, boldface and italics, and regular. Font special effects include such characteristics as underlining, double underlining, superscript and subscript, shadow, outline, emboss, engrave, strikethrough, and double strikethrough. In general, these features are like toggle switches in that you turn them on (or off) by selecting the option once, and you turn them off (or on) by selecting the option a second time. Figure 3-12 illustrates some of the different font styles and effects available in Word.

Figure 3-12
Font styles and effects in Word

This sentence is written in italics.

This sentence is written in bold.

This sentence is written in bold italic.

This sentence is written with underlined text.

This sentence is written with double-underlined text.

~~This sentence is written with the strikethrough text effect.~~

Word's special effects options also include superscripted and subscripted text. The affected text is automatically reduced in size, as illustrated in Figure 3-13.

Figure 3-13
Superscripted and subscripted text

This sentence illustrates ^{superscripted} text.

This sentence illustrates $_{\text{subscripted}}$ text.

To change any or all of the font characteristics in a block of text, first select the text you want to change. Then choose Font from the Format menu. A dialog box similar to the one shown in Figure 3-14 appears. You will use the options in this dialog box to change the character formats in your document. As you make selections within the dialog box, the effect of your changes will be displayed in the Preview window.

Figure 3-14
Font dialog box

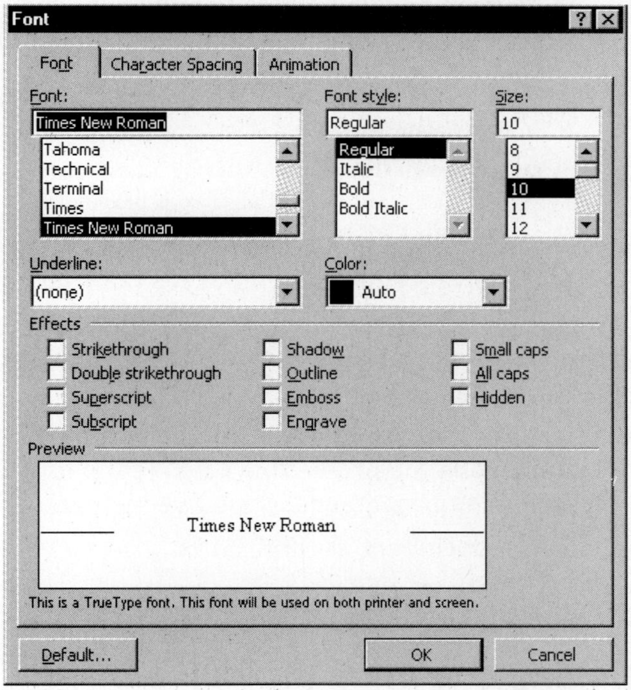

To change a character's font, size, style, or effects, choose the appropriate option. You may need to scroll to find the desired font or size. Single, double, dotted, or other underlining effects can be chosen from the Underline drop-down list box. When you finish customizing text characteristics, choose OK to close the dialog box and apply the changes to your selected text. If you did not select a block of text before you customized the text appearance, the changes you specified will become the default characteristics for the next text you type.

Try It

In this exercise, you customize some of the text in the memo you edited in the last Try It! exercise. The resulting memo should look similar to the following.

MEMO

TO: Instructor's Name

FROM: Your Name

DATE: Current Date

RE: Problem Solving

Engineers solve problems virtually every day of their careers. Some of the problems encountered on the job are relatively trivial, whereas others are extremely complex. You can solve most problems by using the five-step problem-solving method outlined here.

1. **Define the problem.** It is extremely important that you understand the problem before attempting to determine an optimal solution.
2. **Gather the data.** When solving a problem, you frequently start with incomplete data. It is critical to amass all the necessary data before beginning to come up with a solution.
3. **Look at alternative solutions.** Most engineering problems have several possible solutions. You need to consider all the alternatives before committing to a final solution.
4. **Select the "best" solution.** After carefully weighing all the possible solutions, select the solution that satisfies all or most of the design parameters.
5. **Implement the solution.** Virtually all engineered solutions are either constructed or manufactured as the final stage in the problem-solving process.

- Open MEMO2.DOC. Select all the text in your memo, and change the font face to Helvetica. If this font face is not available in your computer lab, choose Courier.
- Select the word MEMO, change the font style to bold, and change the font size to 18.
- Select the short sentence at the beginning of the first problem-solving step ("Define the problem."). Change the font style to bold. Repeat this for each initial sentence in the five-step procedure.
- Save the document as MEMO3.DOC.

Moving and Copying Text

When creating Word documents, you may need to move or copy blocks of text. Moving text erases it from its current location and places it in a new location. Copying text leaves one copy of the text in its original position and places another copy in a new location.

To move a block of text, first select it, and then choose Cut from the Edit menu. The selected text is removed from its original location. Next, move the insertion point to the desired new location, and choose Paste from the Edit menu. The block of text is inserted in its new location.

To copy a block of text, first select it, and then choose Copy from the Edit menu. (Note that the selected text will remain in its original position.) Place the insertion point where you want the copy, and choose Paste from the Edit menu. A copy of the text now appears in the new location.

Aligning the Page Vertically

When you write title pages or short documents such as memos, you may want to center the text between the top and bottom margins. Don't confuse this option with center justification, where each line of text is centered between the left and right margins. To center the text vertically on the page with Word, first choose Page setup from the File menu, and then choose the Layout tab in the dialog box that appears (see Figure 3-3). To choose this tab, point to the word *Layout* at the top of the dialog box (it looks like the tab from a file folder) and click. Figure 3-15 shows the Page Setup dialog box after the Layout tab has been selected.

Three alignment choices are available to you from the Vertical Alignment drop-down list box: Top, Center, and Justified. Top is the default option and is used to align the first line of the first paragraph with the top page margin. Center aligns the paragraphs so that they are centered between the top and bottom margins. Justified spaces the paragraphs evenly so that the top line on a page aligns with the top margin and the bottom of the document aligns with the bottom page margin. From the Apply To drop-down list box, you can choose to apply the vertical alignment to the whole document or from this point forward in the document.

In normal text view, the vertical page alignment will not be displayed. To see the effect of changes to the vertical alignment, choose Print Preview from the File menu. A sample page showing how the document will be displayed will then be shown.

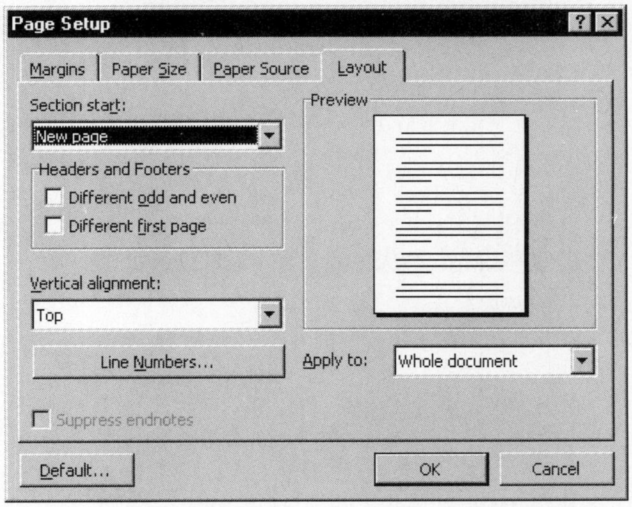

Figure 3-15
Page Setup dialog box with Layout tab revealed

3-4 CHECKING SPELLING AND GRAMMAR

You can use Word to check the grammar and the spelling of your document. However, keep in mind that neither of these tools is infallible. The *spelling checker* does not find mistakes that are words in their own right. For example, if you mean to say *for* but instead you type *four*, the spelling checker does not catch the error because *four* is a correctly spelled word. The *grammar checker* may alert you to an error if a word seems out of place within a document, but it may not.

Checking Your Spelling and Grammar

The Word spelling and grammar tool checks the spelling of all words in your document and checks the grammar of the sentences in your document. Word simultaneously searches your document for either spelling or grammatical mistakes. Although Word searches for both types of errors at the same time, each activity is slightly different and will be described separately. To check the spelling of your document, choose Spelling and Grammar from the Tools menu.

The spelling tool searches Word's standard dictionary for each word in the document and stops on words that are not in this dictionary. Word will automatically start checking the spelling of the words in your document from the insertion point forward. If the insertion point is in the middle of the document when the spelling tool starts, when it reaches the end of the document it will automatically go back to the beginning and resume checking until the entire document is checked.

You use a Spelling dialog box like the one shown in Figure 3-16 to verify your document spelling. In this dialog box, the word *ther* was not found in the dictionary. Beneath this entry is the Suggestions text box; Word guesses words that are spelled similarly to the incorrect word and puts them here. You can scroll through the words in the Suggestions text box to see other possibilities. If one of the words listed here is the word you meant to use, choose the correct word and then choose Change. If you realize that you have misspelled the same word in numerous places within the document, choose Change All to correct all occurrences.

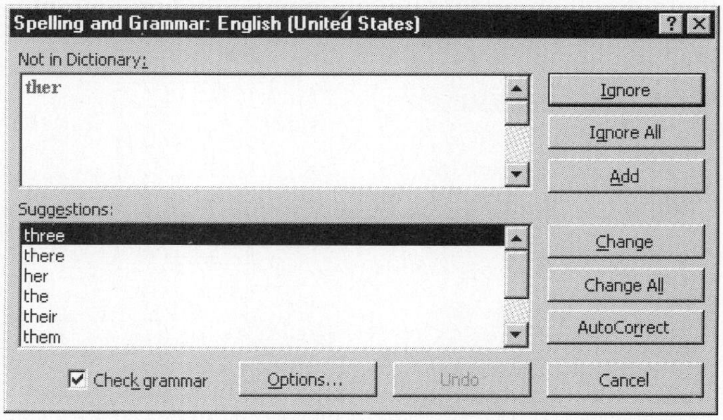

Figure 3-16
Spelling dialog box

When writing technical papers and reports, you will often use words that are not in Word's standard dictionary but that are also not misspellings—for example, terms such as *spalling* or the names of famous scientists like Euler or D'Alembert. In these cases, you can choose Add from the Spelling dialog box to add the word to the Word dictionary. Once you add a word to the dictionary, Word will never again stop on that word during a spell check. If you want to allow a word in the document but do not want to add it to the dictionary, you can choose either Ignore (to allow the word in the document this one time) or Ignore All (to allow the word throughout the document).

Word's grammar checker checks the grammar of your document. If you wish to check only the spelling of your document and not your grammar, clear the Check grammar check box in the Spelling and Grammar dialog box. Some of the types of grammatical errors Word will find include the following:

- Uncapitalized first word in a sentence
- Missing punctuation after phrases
- Plural nouns mixed with singular verbs or vice versa

The grammar checker tells you when it thinks you have used a word incorrectly within the context of the sentence. For example, suppose your document included the following sentence: "The stress was to high." The word *to* should be *too*. If you checked this sentence with the grammar checker, Word would flag the sentence, highlighting the word *to* and displaying the word *too* in the Suggestions box. You could then replace *to* with *too* in the document and continue checking your document's grammar.

Using AutoCorrect

Word's *AutoCorrect* feature automatically corrects mistakes as you type them. For example, if you type "adn" with a space after it, Word will automatically correct it for you and edit your document so that the word now reads "and." You can use this feature to automatically correct for common errors such as using two initial capital letters in a word or to capitalize the first word in a sentence or in the names of days. You can also customize the AutoCorrect feature to replace abbreviated strings of characters. For example, you can use this feature to create an entry so that every time you typed "E" and followed it with a space, Word would replace this with "Modulus of Elasticity." To add an entry for automatic correction, choose AutoCorrect from the Tools menu. A dialog box similar to the one shown in Figure 3-17 appears. You can then make entries in the text boxes labeled Replace and With to customize this feature. In Figure 3-17, an entry has been made to automatically change "E" to "Modulus of Elasticity." If you want to turn the AutoCorrect feature off, clear the Replace Text as You Type check box by clicking on it, and then choose OK.

Figure 3-17
AutoCorrect dialog box

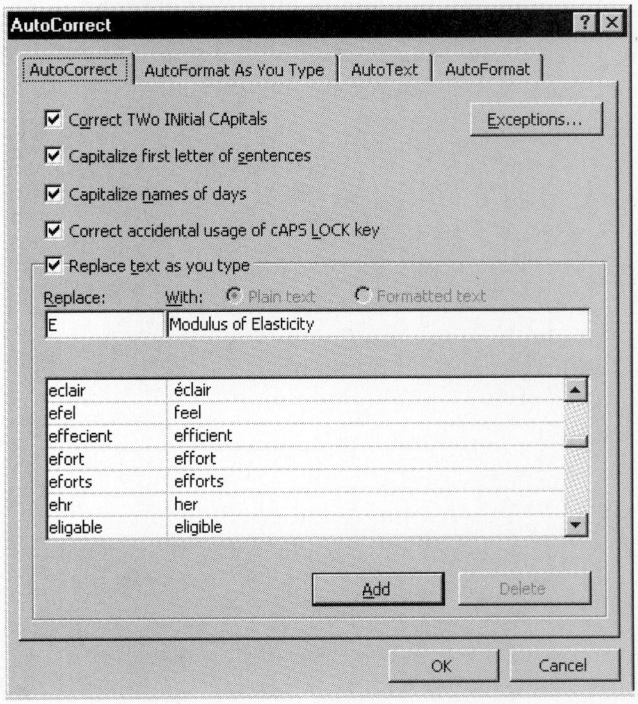

Try It In this exercise, you continue to edit the memo you created earlier.

- Open MEMO3.DOC if it is not already open. Cut the sentence "You can solve most problems by using the five-step problem-solving method outlined here."
- Paste this sentence after the sentence that ends with the word *careers*.
- Center the memo vertically on the page.
- Check the spelling of the document.
- Create a printout of your memo to turn in to your instructor.

3-5 USING SHORTCUTS

You can use the following shortcuts to carry out some of the Word functions described in this chapter:

- **Select text.** Use the mouse to quickly select blocks of text in your document. Double-clicking selects a word; triple-clicking selects an entire paragraph.
- **Cut, copy, and paste.** To display a pop-up menu containing the Cut, Copy, and Paste options, click the right mouse button anywhere in the document. To choose an option from the pop-up menu, click the option name with either mouse button.
- **Set tab stops and margins.** To set tab stops and margins quickly, use the graphical capabilities found on the ruler. You can click the ruler to add tab stops and then drag them across the ruler to their desired location. The default type of tab stop is left, but you can change the type of tab by clicking the button found on the left side of the ruler to change the default (it looks like an L). You can drag existing tab markers off the ruler to delete tab stops.
- **Change indentation.** To change indentation, drag the arrows found on the left and right sides of the ruler. The arrow on the right controls right indentation. The two arrows on the left control left indentation. The top arrow controls the indentation of the first line in the paragraph. The bottom arrow controls the indentation of all paragraph lines except the first. Thus, by changing any of the arrows, all four types of indentation (left, right, hanging, and first) can be set graphically.
- **Use toolbar buttons.** You can also quickly access many Word commands by using the buttons on the toolbars. Click the following buttons to issue the commands outlined in this chapter:

Print preview button

Checks document spelling

Cuts a block of text

Copies a block of text

Pastes a block of text

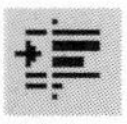
Increases indentation one tab stop to the right

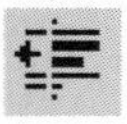
Decreases indentation one tab stop to the left

Sets justified paragraph format

Sets right paragraph justification

Sets centered paragraph justification

Sets left paragraph justification

Underlines text

Italicizes text

Makes text boldface

Sets font size

Sets font

Sets paragraph style

SUMMARY

In this chapter, you learned about the importance of the first stage of the design process: defining the problem. You then learned that letters and memos are the types of documentation most commonly used at this stage. Standard formats for memos and letters were presented. Finally, you learned about a number of Word features that enable you to create customized letters and memos. You discovered how to change margins and tabs; how to modify the paragraph line spacing, indentation, and justification; how to alter several text characteristics such as font, size, and appearance; and how to use Word to check your document's spelling and grammar.

Key Words

AutoCorrect
font
font style
grammar checker
indentation
justification
line spacing
paragraph spacing
points
selecting text
spelling checker
tab stops

Exercises

1. Create a document that shows the Palatino font with seven different font sizes. If you do not have Palatino available, use the Courier font.
2. Create a document that shows ten different fonts, all with the font size of 12 points.
3. Create a document that shows ten different combinations of font size and style.
4. Choose any product currently on the market, and write a memo to your instructor outlining one of its problems from an environmental standpoint.
5. Choose any product currently on the market, and write a memo to your instructor outlining one of its problems from a functionality standpoint.
6. Using the techniques developed in this chapter, create a letterhead for a fictitious company.
7. Write a letter to your instructor outlining one of the most recent problems that you solved using a method other than the five-step approach.
8. Write a letter to your instructor describing a problem and how you would solve it by using the five-step method described in this module.
9. Write a memo to your instructor describing one of the problems you have encountered using Word.
10. Write a letter to your parents or another family member describing what you feel is the most prevalent problem at your university.
11. Write a resume for yourself that includes two different fonts, two different text sizes, and two different text appearances (bold, italics, and so on).
12. Write a letter to someone in a company or business asking about summer employment.

4 Reporting Experimental Data

Application Satellites

Satellites orbit the earth, communicating video and audio information with speed and accuracy. Satellites enable us to follow events as they occur around the world, track threatening weather patterns, map uncharted terrain, and follow volcanic activity. This information is then easily communicated to the general public through televised broadcasts, saving countless lives. Satellites are currently gathering data so that improved methods for predicting long-range weather patterns can be developed. The ability to communicate events as they take place is a distinct characteristic of the second half of the twentieth century. Through satellite communications, we are better informed and better educated than ever before. In this chapter, you will learn about the second stage in the design process (gathering data) and about how engineers communicate experimental results through lab reports.

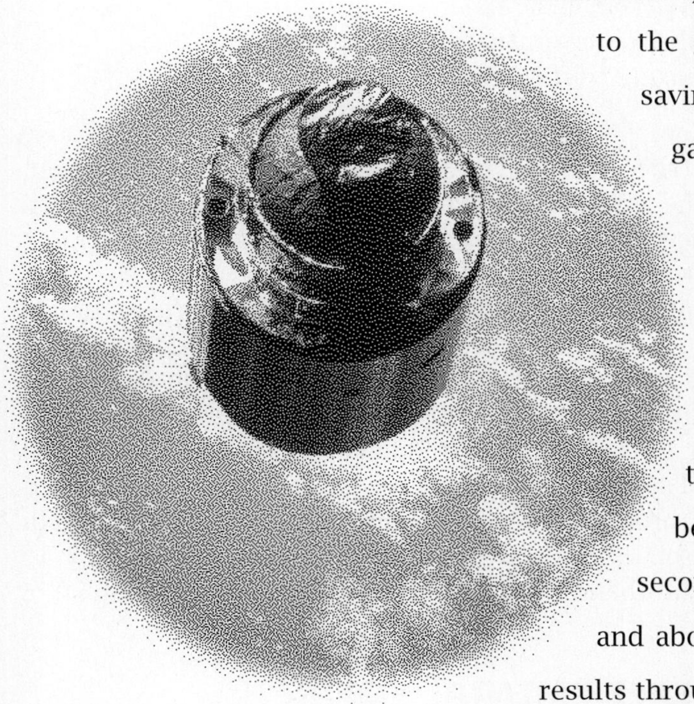

INTRODUCTION

Once you have defined a problem, the next stage of the design process involves gathering data. You could search through books and journals for previous solutions to similar problems. Often, however, you must conduct experimental work in the field or in a laboratory. In this chapter, you learn about reporting experimental data in the form of written lab reports. You will learn about acceptable writing styles for this type of report. You also learn how to add headers and footers to your documents, and you then learn about creating and editing tables within a Word document.

4-1 GATHERING DATA

Gathering data is an integral part of the engineering design process. Gathering data frequently helps you better understand the problem to be solved and can even help you define the real problem more accurately. For example, suppose you are a transportation engineer, and you determine that a certain intersection is unsafe—it is the scene of a high number of accidents and the subject of numerous complaints. Your assignment is to remedy this problem. Before you can solve the problem, however, you must understand its cause.

The factors that influence traffic patterns at this intersection include:

- The major attractions near the intersection
- The number of lanes of traffic
- The driving habits of those using the intersection
- The number and clarity of signs and lane markings posted near the intersection
- The condition of the pavement in the intersection
- The speed limit on the roads approaching the intersection
- The visibility of opposing traffic from the approaches

As the traffic engineer, you need to improve this intersection because people's lives are at stake. However, because you have a limited budget, you cannot rebuild the entire intersection without regard to cost. To determine which of the preceding factors contribute to the problem most significantly, you must go to the scene and study it.

You might gather several different types of data using several methods. You could note personal observations about the traffic flow at the intersection. You could interview motorists and police officers to determine their views on the nature of the problem. You could install car counters to determine the number of cars that go straight or turn right or left at the intersection. You could set up a radar speed study to determine the speed of the vehicles going through the intersection. You could drive through the intersection yourself at several different times during the day to observe the problems firsthand.

4-2 DESIGNING THE EXPERIMENTAL REPORT

After finishing your study of the intersection, you prepare a report outlining the data that you gathered. This document is called an *experimental report.* As an engineering student, you will conduct laboratory experiments and will then most likely be required to submit experimental lab reports summarizing your results. The remainder of this chapter explains how to write experimental reports using Word.

As an engineering student and as a practicing engineer, you will write and read numerous lab reports throughout your career. Much of your undergraduate writing will probably be in the form of experimental lab reports. When writing experimental reports, you should follow a few general rules and should also check with your instructors to see if they have any of their own unwritten rules or preferences.

- **Avoid using the first person.** In other words, do not use words such as *I, me, we,* or *us.* The first person is generally avoided in any type of technical documentation.
- **Use the passive voice.** For example, instead of writing something like "We weighed 50 grams of compound A and put it in a beaker," it is usually preferable to write "50 grams of compound A were weighed and placed in a beaker." Note that in the passive voice it is not apparent who (or what) is performing the action. Most composition instructors frown on the use of the passive voice, and Word's grammar checker will even alert you to the presence of the passive voice and suggest that you change sentences from the passive to the active voice. For most technical reports, however, the passive voice is preferred, although this may be beginning to change. Note that you may sometimes find it necessary to use the active voice for the sake of clarity. (Again, check with your instructor about his or her preferences.)
- **Be concise.** The quality of the report is much more important than its length. Make sure that your report includes all the necessary information without being repetitive or belaboring any points.
- **Check spelling and grammar.** As you learned in the previous chapter, Word provides tools for checking spelling and grammar. Take advantage of them. At the same time, you should proofread the report to check for errors that the spelling and grammar tools may have missed.
- **Compare your results with reference values.** Many of the properties and results that you obtain experimentally will have corresponding values in books or articles. For example, if you experimentally determine the density of a given material, you should look up the density in a reference book for comparison purposes. When you compare your results with *reference values,* you should state where they came from as well as the percent deviation between your results and the reference values.
- **Explain any results that do not correspond well with the reference values.** Do not list "experimental error" as the only reason for the deviation. Try to list some of the sources of this experimental error.
- **Use tables or graphs to display large amounts of data.** Try not to bury the results within the text of the report. Using tables or graphs can make your data easy to locate and easy to grasp.
- **Include sample calculations to illustrate key points.** Sometimes it is best to include calculations as an appendix to the report.

In certain cases, you may need to break some or all of these rules. Remember that they are general rules for preparing technical reports. Before you begin to write, you should check with your instructor about her or his preferred writing style, and you should also consider your audience.

Experimental Report Format

Usually, you divide experimental reports into a number of sections, each of which is preceded by a heading. You can customize the headings so that they are bold, or underlined, or of a larger font size than the body of the report. Here is an outline of the sections you might include in a typical experimental report.

Introduction — This section contains a short introductory paragraph that states the purpose of the experiment. In some instances, this section summarizes the experimental results to make the report easier to follow.

Procedure — This section describes the experimental procedure that was followed. Typically, you use the passive voice and do not use the first person in this section of the report.

Basic Theory — In certain types of reports, this section is optional. It may include theories from books or journal articles, and it may include equations describing the theory. (You will learn how to write equations with Word in Chapter 5.)

Data and Results — If you have a great deal of experimental data, tabulate it in a logical manner so that readers can understand it quickly and easily. Sometimes it is preferable also to include a graph to show trends in the data. When referring to a table of data within an experimental report, make sure that the rows and columns of the table are labeled and that the table itself has a title.

Conclusions — In this section, you state any conclusions that can be drawn from the experimental data. You also compare the experimentally determined values with the reference values and state the source of the reference values. This section should mention any sources of experimental error in the data and any problems you had performing the experimental work.

4-3 CREATING EXPERIMENTAL REPORTS IN WORD

Experimental reports typically involve the presentation of large quantities of data. Tables and graphs are the two most popular methods of presenting data. Graphing data is beyond the scope of this text. This section focuses on creating tables of data. Of course, you also use the formatting techniques you learned in Chapter 3 when creating experimental reports.

Figure 4-1
Sample page with headers and footers

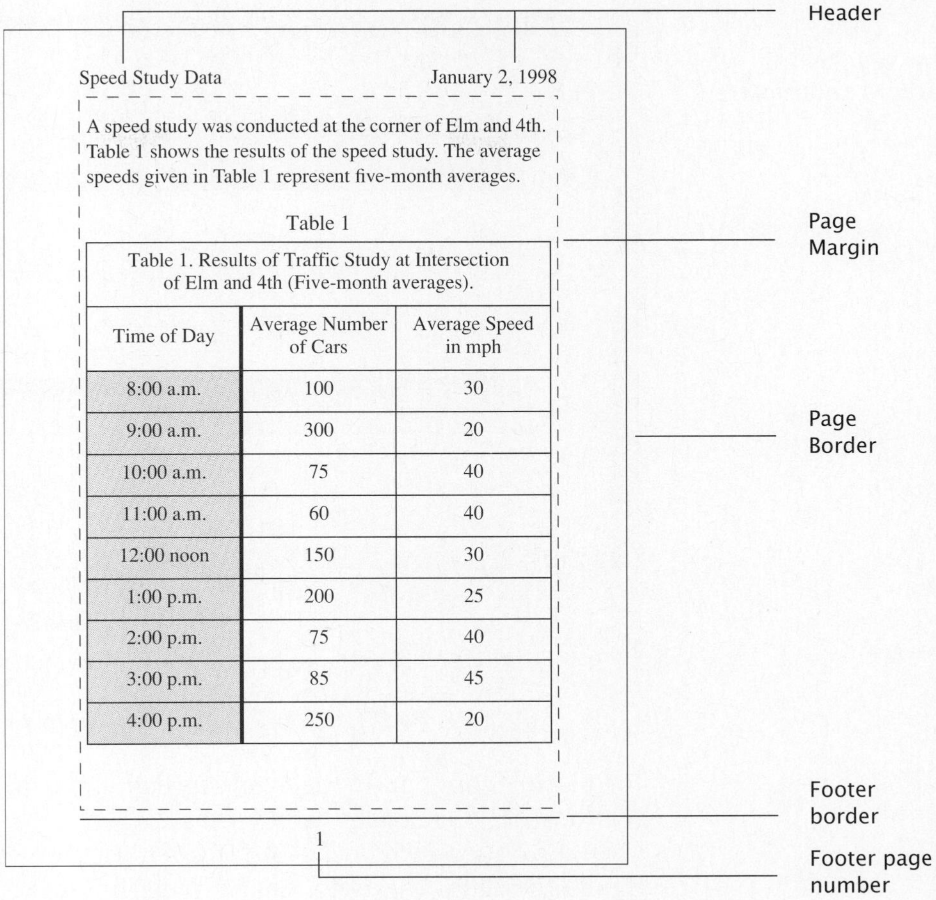

Table 1. Results of Traffic Study at Intersection of Elm and 4th (Five-month averages).		
Time of Day	Average Number of Cars	Average Speed in mph
8:00 a.m.	100	30
9:00 a.m.	300	20
10:00 a.m.	75	40
11:00 a.m.	60	40
12:00 noon	150	30
1:00 p.m.	200	25
2:00 p.m.	75	40
3:00 p.m.	85	45
4:00 p.m.	250	20

Including Headers and Footers

Headers and *footers* are short blocks of text at the top and the bottom of each page in a document. Word automatically places headers and footers created by you in the top and bottom margins of your document. Document headers and footers usually include such information as the page number, the date, the chapter number, or the section heading. Figure 4-1 shows a document that includes headers and footers.

Word documents include headers and footers in separate text boxes in the top and bottom page margins. Normally, when you are working in your document you will not see the header and footer text unless you view the document in print preview or in page layout.

To include headers or footers in your Word document, choose Header and Footer from the View menu. The page layout view of the document will change appearance, a Header text box appears at the top of the current page of your document, and the Header and Footer toolbar appears. This is illustrated in Figure 4-2. (Note: You may have to drag the Header and Footer toolbar to move it into the position shown in Figure 4-2.) The Header text box is automatically set with a center tab stop in the middle of the page and a right tab stop at the right margin of the page. You are able to enter approximately three lines of text within this Header text box. To include text as a Header, type it inside this text box.

If you want to include a Footer in your document, click the button to switch between the Header and Footer text boxes (this button is on the Header and Footer toolbar). You can use any of the font and paragraph

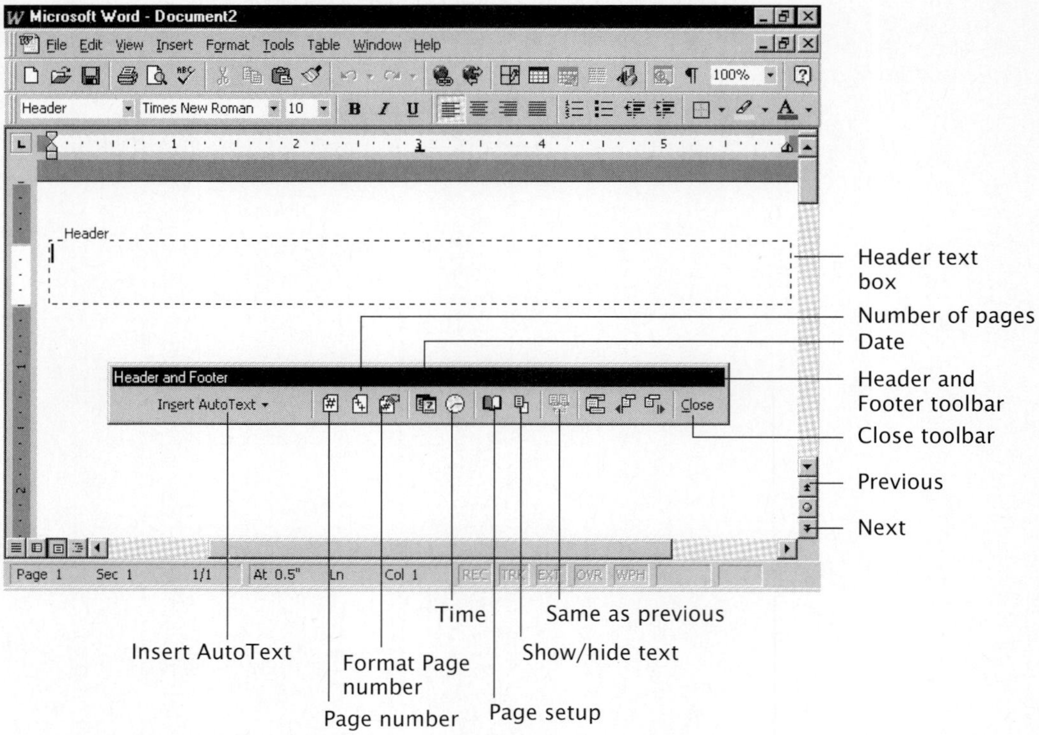

Figure 4-2
Page display for headers and footers

formatting commands that you learned about in Chapter 3 to customize the look of your headers and footers.

Other options found on the Header and Footer toolbar can be used to insert the date, the time, or the page number as part of the header or footer. Some of the options in the Insert AutoText menu include adding your name, the page number and the date, adding text such as *Created by* followed by your name, Page X of Y, and so on. When adding headers or footers with the AutoText options their locations will automatically be set for you. For example, if you choose to add the Author name, Page Number, and Date, the author name will appear at the left margin, the page number will be centered, and the date will be at the right margin.

When you are finished creating headers and footers in your document, click the Close button on the toolbar to return to editing your document. In Normal page view, the headers and footers are not visible as you type.

If you want to further customize your headers and footers, use the options on the Page Setup dialog box (choose Page Setup from the File menu). Here you can set the spacing between the edge of the paper and the Header and Footer text boxes. Selecting the Layout tab from this dialog box enables you to set up your document so that odd and even pages have different headers and footers or so that the first page has a different type of header and footer.

Adding Borders and Shading

When creating documents with Word, you may wish to include borders around certain paragraphs or to add background shading. In particular, you may want to include a border below the bottom of the Header text box or above the Footer text box to separate the headers and footers from the remainder of the text on the page.

To include a border in your document, first place the insertion point in the paragraph that you wish to have a border, or select the block of text for a border and then choose Borders and Shading from the Format menu. You can also make this menu selection while working within the Header or Footer text boxes. A dialog box similar to the one shown in Figure 4-3 appears. Click the Border preview area on the side where you want a border. To change the line style of the border, click the new style from the appropriate list box. You can also use the options in this dialog box to set the distance between the border and the text.

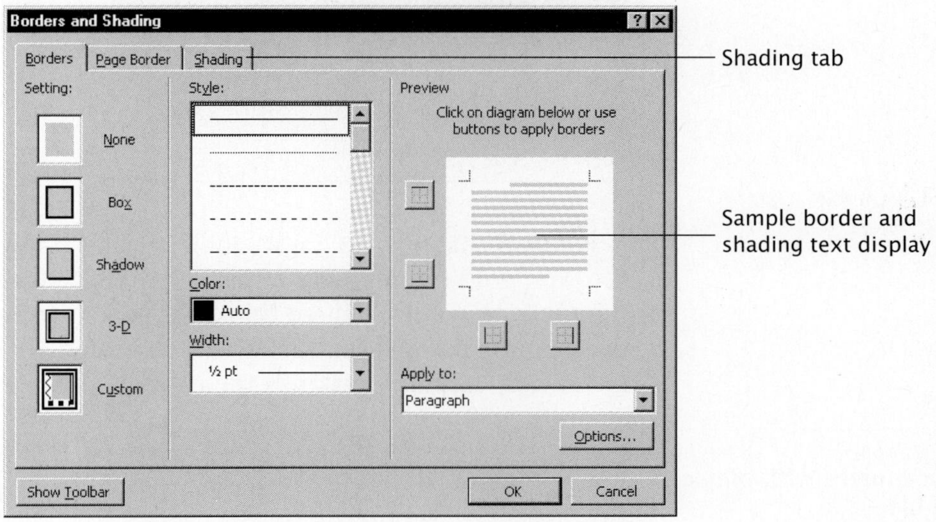

Figure 4-3
Borders and Shading dialog box

To add shading to a paragraph, click the Shading tab in the Borders and Shading dialog box. You can then specify the type of shading and colors you wish to have within the selected paragraphs.

Try It

In this exercise, you create headers and footers in a Word document.

- Start Word if you have not already done so.
- Create a header for your document that contains your name at the left margin and the current date at the right margin.
- Create a border at the bottom of the header that consists of a double thin line.
- Create a footer for your document that consists of the page number centered between the right and left margins.
- Press ENTER enough times until you start a second page. Use the Print Preview command to verify that Word inserted the headers and footers on both pages and numbered the pages automatically.

Creating a Table

Tables are a good way to display large quantities of numbers or text; they are especially useful for displaying data obtained in experimental work. Word makes it relatively easy to create and customize tables. When you add a table to a document, make sure you give it a title, and clearly label rows and columns. This helps the reader identify and understand the data.

To create a table, choose Insert Table from the Table menu. The dialog box shown in Figure 4-4 appears. This dialog box permits you to create a table with an unlimited number of *rows* and up to 31 *columns.* When creating a table, you should include extra rows and columns for labels and a title. The *table title* describes the data displayed in the table. You should also *label* the table columns and rows with a short descriptive name of their contents.

Figure 4-4
Insert Table dialog box

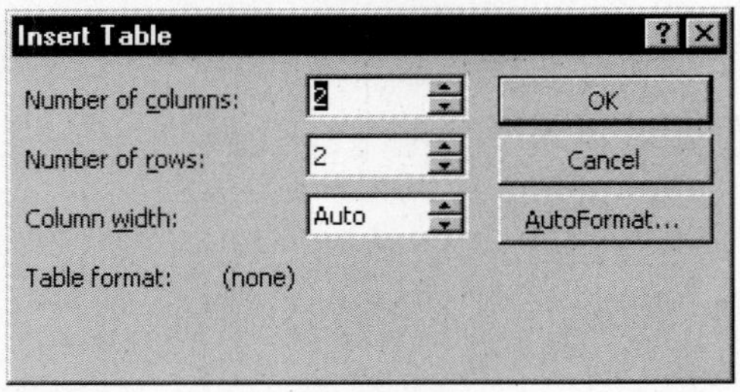

Figure 4-5 shows an empty table with five rows and four columns. Each row and column intersection is called a *cell.* By default, the lines defining the table cells will print as thin lines as shown in the document. Note also the change in appearance of the Word ruler. Each table column is defined by its own column border in the ruler.

Figure 4-5
Document with blank table

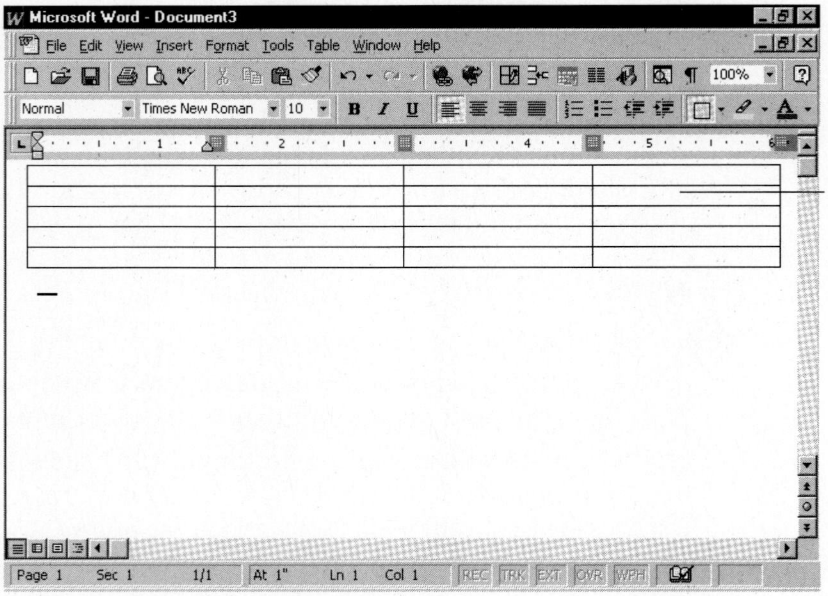

Entering Values in a Table

To enter a numerical or text value within a table cell, place the insertion point in that cell and enter the desired text. To move to other cells in the table, click within the cell, use the arrow keys, or press (TAB) to move from left to right.

You can also perform certain spreadsheet-style operations within Word tables such as add; subtract; multiply; divide; or compute averages, percentages, and minimum and maximum values. These operations are similar to those performed in most standard spreadsheets and are beyond the scope of this text. For details, consult the online Help.

Try It

In this exercise, you create a simple table and save it in a file. When you are finished, your table should look similar to the one shown here:

Time of Day	Average Number of Cars	Average Speed in mph
8:00 a.m.	100	30
9:00 a.m.	300	20
10:00 a.m.	75	40
11:00 a.m.	60	40
12:00 noon	150	30
1:00 p.m.	200	25
2:00 p.m.	75	40
3:00 p.m.	85	45
4:00 p.m.	250	20

- Open a new document. Create a table with 11 rows and 3 columns.
- Place the insertion point in the cell defined by the first column and the second row of the table. Type the column label Time of Day.
- Using TAB or →, move the insertion point around the table, and enter all cell values.
- Save the table in a file called TABLE1.DOC.

Editing a Table

Once you have created a table, you can customize its appearance much as you customize ordinary text. In the following sections, you will learn about some of the more useful table editing features. The more advanced features will be left to you to discover on your own, using online Help. When editing your table, you can choose to edit a single cell or the entire table. You select cells, rows, and columns as you selected blocks of text. Point to the first cell to be modified, and drag across the desired cells to select them. You can then execute the desired editing commands.

Formatting the Contents of Cells Table cells are formatted just like paragraphs are in Word. First, select a single cell or a row or a column of cells; then apply any of the standard formatting commands to them. For example, if you choose Paragraph from the Format menu, you can change the paragraph alignment so that the cell contents will be center-justified or right-justified. You can also set indentations for the cell contents similar to the way that you set paragraph indentations earlier. By selecting Font from the Format menu, you can edit the font, text size, or text style of the cell contents, similar to the way that you edit the font characteristics in your document.

Splitting or Merging Cells Sometimes you may need to split or merge columns or rows in a table—especially when labeling columns that contain subsets of data or when creating a table title.

To split a table cell, place the insertion point in the cell and choose Split Cells from the Table menu. Next, indicate how many columns you wish to split the original cell into. (The default is 2.) You can also split an entire column by selecting that column before issuing the Split Cells command. Figure 4-6 shows the effect of splitting the last four rows of a column into three subcolumns. (Note that the first two rows were not split.)

Figure 4-6
Splitting cells

Time of Day	Speed in mph		
	Point 1	Point 2	Point 3
8:00 a.m.	25	20	30
1:00 p.m.	25	35	30
5:00 p.m.	15	25	25

Merging table cells simply means combining them. Cells you wish to merge must be in the same row or column of the table. To merge cells, select the cells to be merged and choose Merge Cells from the Table menu. Figure 4-7 shows the results of merging the two columns in the first row of the table shown in Figure 4-6 and then inserting a table title in the resulting single cell.

Figure 4-7
Merging cells

Results of Speed Study			
Time of Day	Speed in mph		
	Point 1	Point 2	Point 3
8:00 a.m.	25	20	30
1:00 p.m.	25	35	30
5:00 p.m.	15	25	25

Try It

In this exercise, you customize the appearance of the table you created in the previous Try It! exercise.

- Open TABLE1.DOC. Center-justify the two columns labeled Average Number of Cars and Average Speed in mph.
- Change the font of the labels to bold and italic.
- Select the cells in the first row of the table, and merge them into a single cell.
- In this title cell, type Table 4-1. Results of Traffic Study at Intersection of Elm and 4th (Five-Month Averages).

- Edit the table title so that it is center-justified, bold, and has a font size of 18.
- Save the table in a file called TABLE2.DOC.

Customizing the Appearance of Cells When creating tables, you can customize the cell borders and shading (or fill) of selected cells in much the same way that you add borders and shading to text paragraphs in the document. Both of these options draw the reader's attention to important data within the table. To customize the shading or borders in your table, first select the cells to be changed, then choose Borders and Shading from the Format menu.

To customize a border, first select the cell or cells that share that common border, and then choose Borders and Shading from the Format menu. You then click the line style you want for that one particular border (scroll through the selections, if necessary), and click the border you wish to change in the Border preview area.

To customize the cell shading, select the cells you wish to edit. Choose the Shading tab from the Paragraph Borders and Shading dialog box, and then apply the desired shading to the cells. In Figure 4-8, some of the cell borders were customized, and shading was added to accentuate the slowest speeds at each sampling time from the speed study.

Figure 4-8
Customizing borders and shading

Results of Speed Study			
Time of Day	Speed in mph		
	Point 1	Point 2	Point 3
8:00 a.m.	25	20	30
1:00 p.m.	25	35	30
5:00 p.m.	15	25	25

Changing Column Widths When you create a table, by default Word automatically adjusts column widths so that the table fits between the right and left page margins. Sometimes, however, you need to adjust the width of selected columns to make the text fit well within their borders. You can drag the column borders to adjust column widths. First, point to the border you wish to move and hold down the mouse button. The mouse pointer changes into a double-headed arrow, and a dashed vertical line representing the column border appears, as shown in Figure 4-9. You then drag this column border to the desired location and release the mouse button to change the width of the column.

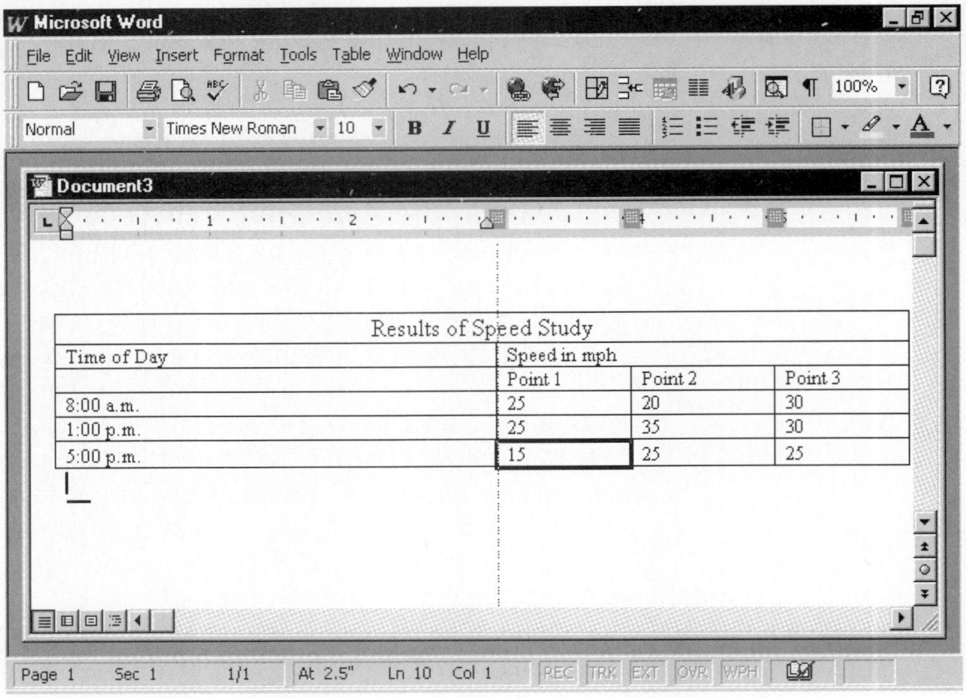

Figure 4-9
Changing column widths

Try It In this exercise, you change the fill and line styles in the table you have been working on. When you are done, your table should look similar to the one shown here.

Table 1. Results of Traffic Study at Intersection of Elm and 4th (Five-Month Averages)		
Time of Day	***Average Number of Cars***	***Average Speed in mph***
8:00 a.m.	100	30
9:00 a.m.	300	20
10:00 a.m.	75	40
11:00 a.m.	60	40
12:00 noon	150	30
1:00 p.m.	200	25
2:00 p.m.	75	40
3:00 p.m.	85	45
4:00 p.m.	250	20

- Open TABLE2.DOC. Place the insertion point in the cell containing the table title. Change the line style for the bottom border of this cell to a double thin line.
- Select the Time of Day column (but do not include the cell with the label). Change the line width of the right border to a thick single line (2 or 3 points is fairly thick), and change the shading to 10% fill.
- Decrease the width of the leftmost column so that it is just wide enough for the column label (Time of Day).
- Save the document as TABLE3.DOC.

Inserting or Deleting Cells When editing a table, you may need to add or delete rows or columns. You may wish to delete entire rows or columns from the table, or you may just want to remove the cell contents without removing the cells themselves. To delete the contents of table cells, select the cells in question. Then press DEL from the keyboard.

To delete the cells themselves, select the cells you wish to delete, and then choose Delete Cells from the Table menu. A dialog box like the one in Figure 4-10 appears. From this dialog box, if you choose the option Shift Cells Left, then after the selected cells are deleted, anything that is currently to the right of the cells will be shifted to take the place of the deleted cells. Similarly, if you select Shift Cells Up, cells that are currently below the selected cells will be shifted up to take their place. You could also choose to delete the entire row or column that contains the selected cells.

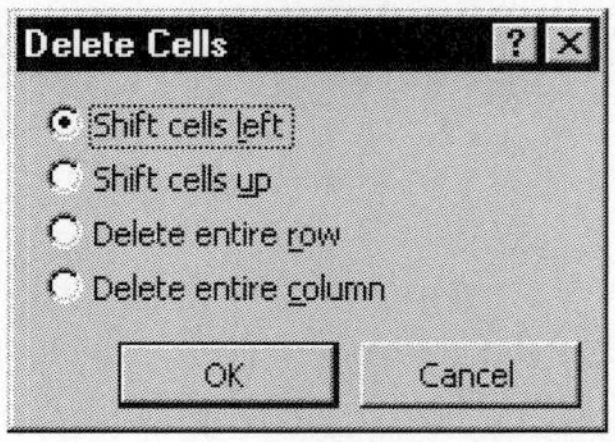

Figure 4-10
Delete Cells dialog box

To add cells to your table, first select the cell or cells at the location where you wish to insert cells, and then choose Insert Cells from the Table menu. A dialog box like the one shown in Figure 4-11 appears. If you select the option Shift Cells Right, blank cells are inserted at the current location, and all other cells in those rows shift to the right. If Shift Cells Down is selected, new cells are inserted at the current location, and all other cells in that column are shifted down. You can also choose Insert Entire Row to add a row above the current location or Insert Entire Column to add a column to the left of the current location. If you place the

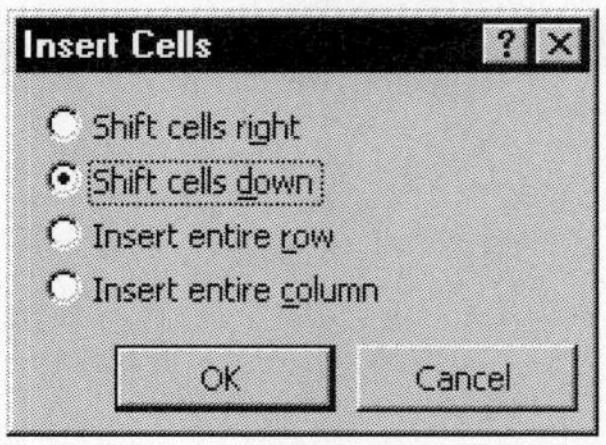

Figure 4-11
Insert Cells dialog box

insertion point inside a cell instead of selecting the entire cell, you can choose Insert Row from the Table menu to insert an entire row above the insertion point location. To insert rows at the bottom of a table, place the insertion point in the last cell of the table and press the TAB key on the keyboard.

Try It In this exercise, you continue to customize your table so that it looks like the one shown here.

Table 1. Results of Traffic Study at Intersection of Elm and 4th (Five-Month Averages)

	Average Number of Cars	*Average Speed in mph*
8:00 a.m.	100	30
9:00 a.m.	300	20
12:00 noon	150	30
1:00 p.m.	200	25
2:00 p.m.	75	40
3:00 p.m.	85	45
4:00 p.m.	250	20
5:00 p.m	200	30
6:00 p.m.	100	45

- Open TABLE3.DOC. Insert two rows after the last row of the table. In these two new rows, enter the following data:

Time of Day	Average Number of Cars	Average Speed in mph
5:00 p.m.	200	30
6:00 p.m.	100	45

- Because the hours of 10:00 a.m. and 11:00 a.m. have very little traffic, delete these two rows from your table.
- Because the column label Time of Day is obvious, delete the contents of this cell.
- Save your file under the filename TABLE4.DOC. Obtain a printout to turn in to your instructor.

4-4 USING SHORTCUTS

Most of the shortcuts you used to customize documents can also be used to create or edit tables. In addition, a few toolbar buttons are useful for creating tables:

Inserts a table (when you are working on a table, this button can be used to insert rows)

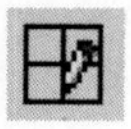

Adds the Borders and Tables toolbar (shown in Figure 4-12)

Figure 4-12
Borders and Shading toolbar

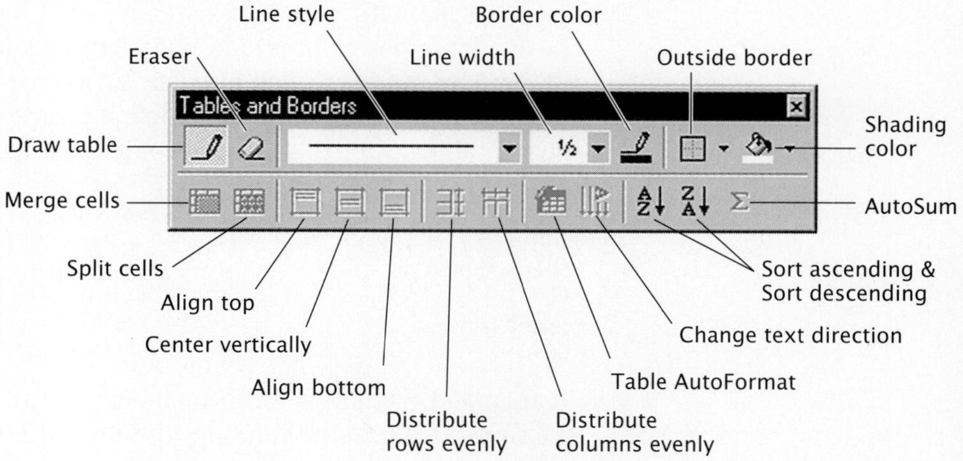

SUMMARY

In this chapter, you learned about the importance of gathering data, which is the second stage of the engineering design process. You learned the proper format for experimental reports. You also learned about some key Word features that enable you to create experimental reports, such as how to add headers and footers to your document, how to add page numbers to headers or footers, how to create tables for the display of experimental data, and how to customize the appearance of tables.

Key Words

cell	label
column	reference values
experimental reports	row
footer	table
header	table title

Exercises

1. Using the header and footer options, create a sample letterhead for a fictitious company.
2. Create the table shown here. Include a centered table title.

Data Point	Change in Time (sec)	Change in Distance (mm)	Velocity (mm/sec)
1	1.5	315	210
2	1.4	300	214
3	1.7	325	191
4	1.6	320	200
5	1.4	305	218

3. For an airplane to maintain a steady, level flight for a given airspeed, the available power provided by its engines must exceed the required power. The power available to the airplane is also a function of its airspeed, because the propeller efficiency varies with airspeed. The following table supplies data for a particular aircraft:

Airspeed (mph)	60	70	80	90	100	110	120	130	140	150
Power Required (HP)	53	51	52	55	61	70	82	96	113	134
Power Available (HP)	90	99	105	110	114	117	120	121	122	121

Rearrange this data so that the labels identify the columns instead of the rows. Use custom shading to indicate the maximum safe airspeed.

4. The following data is from a sieve analysis for an aggregate to be used in a concrete mixture. Create the table, and enter all values. Modify the table to include custom cell borders and shading. Note that the percent retained is the weight retained on the sieve divided by the total weight of the aggregate. In this case, the total weight is 500.0 gr.

Sieve No.	Sieve Opening (mm)	Weight Retained on the Sieve (gr)	Percent of Weight Retained on the Sieve
3/8"	9.5	0.0	0
4	4.75	8.5	42.3
8	2.36	68.1	
16	1.18	130.8	
30	0.600	105.2	
50	0.300	72.1	
100	0.150	61.6	
Pan		19.9	

5. The following data was taken from a compaction test of a brown silty clay. Create the table, enter the data provided here, and fill in the blank cells (use your calculator). The moisture content is defined as the ratio of the weight of water to the weight of solids. One sample calculation has been done for you. All weights are given in grams. Include an extra row at the bottom of your table that shows the average moisture content.

Moisture Content Determination

Test No.	Can No.	Weight of Wet Soil and Can	Weight of Dry Soil and Can	Weight of Water	Weight of Can	Weight of Dry Soil	Moisture Content
1	A	235.65	210.38	25.27	45.20	165.18	15.3%
2	B	231.50	203.55		43.83		
3	C	236.48	203.27		43.14		
4	D	256.63	217.16		44.80		
5	E	243.99	204.07		42.89		
6	F	244.61	201.81		43.61		
7	G	239.87	196.11		42.88		
8	H	251.01	202.75		44.20		

6. Flip a coin 50 times and record the number of times the coin turned up heads and the number of times it turned up tails. Using headers and including page numbers in the footer, write an experimental report about this experiment. Compare your results with theoretically expected values, and tabulate your data.
7. Conduct an experiment similar to the one outlined in Exercise 6. This time, however, roll a single die or a pair of dice, record the results, and write a report outlining the experiment.
8. Conduct an experiment in a building at your college (maybe the cafeteria or student union). For the people who come through a given entrance, count the number who turn right, turn left, go up- or downstairs, and so on during a busy ten-minute period. Write an experimental report that shows the data in tabular form.
9. Open TABLE4.DOC, which you created in one of the Try It! exercises in this chapter. Add a column that displays the average car speed in units of kilometers per second (km/sec).
10. Create a Word table that shows your weekly schedule. Include time for studying, eating, sleeping, attending class, and goofing off.
11. Go to the library and search old newspapers to determine the high, low, and average temperature in your city for the 15th of each month during the past year. Report your findings in tabular form.
12. Go to the library and find a book on technical writing. Write a summary of its instructions for writing lab reports.

5 Writing Proposals

Fiber-Optic Communications

Fiber-optic cables, which carry more information at higher speed than traditional copper wires, have vastly improved our ability to communicate with people around the world. One of the primary advantages of fiber-optic communication is that the signal is transmitted with virtually no noise, resulting in a clearer signal. On the downside, fiber-optic cables are generally more expensive than copper wires. Trade-offs like these are often considered when engineers are evaluating several alternative design solutions, during the third stage of the design process. This chapter explores that stage and explains how to develop proposals to investigate design alternatives.

INTRODUCTION

Most engineering problems have many possible solutions. Engineers must weigh several alternatives before deciding which one comes closest to the design criteria. In this chapter, you learn about the third stage in the engineering design process: investigating alternatives. You also learn how to write proposals for conducting technical investigations. Proposals often rely on the work of previous investigators who have quantified their results in an equation. Thus, equations are frequently used in writing engineering proposals. In this chapter, you also learn how to create and edit equations in Word.

5-1 INVESTIGATING ALTERNATIVES

In the third stage of the design process, engineers look at several possible solutions to the problem at hand. In weighing the alternatives, you must consider factors such as cost, functionality, life cycle, maintenance requirements, aesthetics, and the client's wishes. Typically, each of the possible solutions will meet or exceed the minimum standards in a few of these areas but not in the others. Sometimes, one solution does well in all areas, but this is rare. The engineer's job is to carefully consider each alternative solution and determine the most important criteria before deciding on the final design.

For example, suppose you are an environmental engineer and part of a team responsible for the environmental impact of manufacturing processes at a large corporation. In this capacity, you would probably examine each manufacturing process to see how it affects the other processes and the environment. Frequently, you may have to choose between improving the processes so that fewer pollutants are created or cleaning up the pollutants that are created from the existing manufacturing processes. Some of the factors you may consider in weighing the alternatives include the following:

- The cost of changing the manufacturing process so that fewer waste materials are produced
- Whether the type of pollutant being produced is highly toxic
- Whether the pollutants are treatable and, if so, whether the treatment will be cost-effective
- Whether you have access to a disposal facility that you can use to get rid of the waste material economically
- Public perception of the corporation
- Government regulations regarding environmental protection
- Future liability to the company
- Your own conscience as a world citizen

As a part of the team reviewing possible solutions to the problem, you may be asked to investigate one specific solution—for example, the use of underground storage tanks for a particular pollutant. You must consider the type of soil where the tanks will be located, the toxicity of the residual, the proximity of the water supply source, governing federal and state regulations, and whether these storage tanks represent a viable long-term solution to the problem (presumably, you have a finite amount of space available for this type of storage). As a part of this activity, you might be asked to write a proposal detailing your investigation of some of the issues surrounding the use of underground storage tanks. Other members of your group may write proposals regarding investigations about other potential solutions to the problem. The results of these investigations would then be used to help choose the final design solution.

5-2 PROPOSAL FORMAT

A *proposal* is a detailed plan for an investigation of a design alternative. Proposals can contain provisions for conducting experimental work as well as the development of basic theory. Before you write a proposal, you should have a well-defined plan of action that is likely to bring about the desired results. The writing style for a proposal is similar to that for an experimental report. However, you can sometimes use the first person in a proposal. Here is an outline for a typical proposal.

Introduction	In this section, you thoroughly define the problem and the need for the investigation.
Results from previous work or background	If others have investigated a similar problem, you should present their results in this section. Typically, this section is a compilation of the results from previous work as reported in books and journals. Most likely, you will need to do research to find the information to be included in this section of the proposal. (You will learn how to cite references in the next subsection of this chapter.)
Objectives	This section of the proposal describes the project goals and the results you hope to obtain by carrying out the investigation.
Project description	In this section, you outline the experimental, numerical, or theoretical methods you will use to conduct the investigation. You also describe what equipment or software is to be used in the proposed work.
Future work	This section is optional. It should outline where you think the proposed work will lead—that is, what you think the next step in the investigation will be after the current project is completed. Other related projects or investigations may sometimes develop during the proposed investigation.
Personnel and budget	This section details who will conduct the investigation and how many person-hours will be required for the work. It also outlines the cost of purchasing equipment or supplies for the proposed project.
References	In this section, list all the journal articles and books cited in the proposal.

Citing References

In most technical writing, you must credit authors for their ideas by citing their work. In fact, it is highly unethical to use someone else's ideas or to copy sections of text in a proposal or a paper without giving proper credit. In general, a proposal can include figures and equations from other sources as long as you cite the reference. There are several acceptable ways of citing references in proposals; we will focus on two of them.

List all the papers and books cited in your proposal at the end of the proposal in the reference section. You can list the references either in alphabetical order or in the order in which they appear in the proposal. Find out the preferred citation method for your situation before writing your proposal.

When listing references in the order in which they appear in the proposal, you should include the reference number in the document text, enclosed within parentheses or brackets. The first reference will be [1], the second [2], and so forth. In the reference section, list each reference that corresponds to the numbers as cited. In the reference list that appears at the end of the proposal, list the items consecutively by number.

If you list references alphabetically by the author's last name, include the author's name and the publication year in parentheses within the document text. A correct citation in this format looks like this: (Duits, 1993).

Reference formats vary depending on whether the work is a book or an article. The format for citing books is as follows:

Earle, J. H., *Engineering Design Graphics,* Addison-Wesley,
Reading, Massachusetts, 1987, pp. 264–265.

When citing journal articles, the following format is widely used:

Nusholtz, G. S., J. Wu, and P. Kaiker, "Passenger Air-bag Study Using
Geometric Analysis of Rigid-body Motion," *Experimental Mechanics,*
Vol. **31**, No. **3**, September 1991, pp. 264–270.

Note that each type of reference includes the author's name, the title of the work, the publisher, and the date of publication. For articles, include the volume and number of the journal. In general, a reference should include enough information so that someone reading your proposal can locate the work in question.

Sometimes you will be required to use footnotes to cite references within your document. In Word, you can automatically create footnotes at the bottom of the current page or endnotes at the end of the document. To create a footnote or endnote, choose Footnote from the Insert menu. From the dialog box that appears, you can select either Footnote or Endnote. You can also choose to have Word automatically number the footnotes for you or to customize the numbering scheme. Word will then insert the number in superscripted form in the document and jump to the footnote location so that you can then insert the footnote information at the bottom of the page.

5-3 CREATING PROPOSALS IN WORD

You can use all of the Word features you have already learned about when creating proposals. For example, proposals often include tables, which were discussed in detail in Chapter 4. Proposals typically contain a section on background material, so they commonly contain equations. Your proposal could contain a statement similar to the following:

It is well established that the stress-strain relationship for a linear-elastic member in uniaxial tension is defined by the following:

$$\sigma = E\varepsilon \qquad \text{Eq. (1)}$$

where σ is the stress, E is Young's module, and ε is the strain. The project described in this proposal will investigate the applicability of this relationship to other, nontraditional materials.

The remainder of this chapter explains how to create and modify equations using Word.

Creating Equations

Word enables you to create equations within your document and to customize the appearance of these equations. Note that Word is used only to insert the equation into your document; it cannot be used to solve mathematical problems.

To create an equation, choose Object from the Insert menu. A dialog box like the one shown in Figure 5-1 appears.

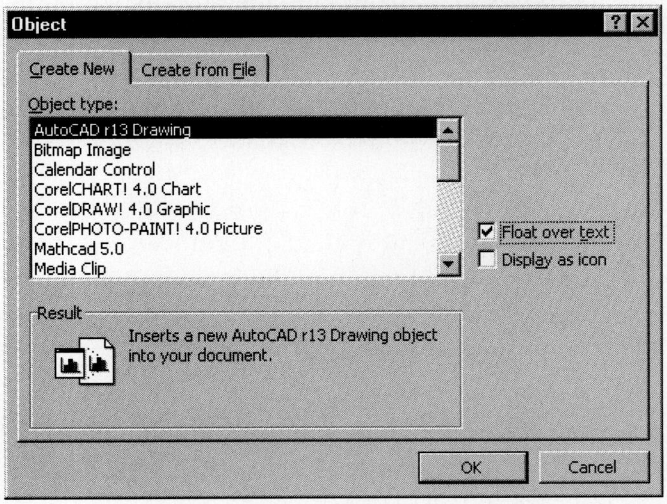

Figure 5-1
Object dialog box

To create an equation at the insertion point in your document, select Microsoft Equation 2.0 from the Object Type list box. When this option is selected, the screen changes appearance and is similar to that shown in Figure 5-2.

Figure 5-2
Equation Editor screen layout

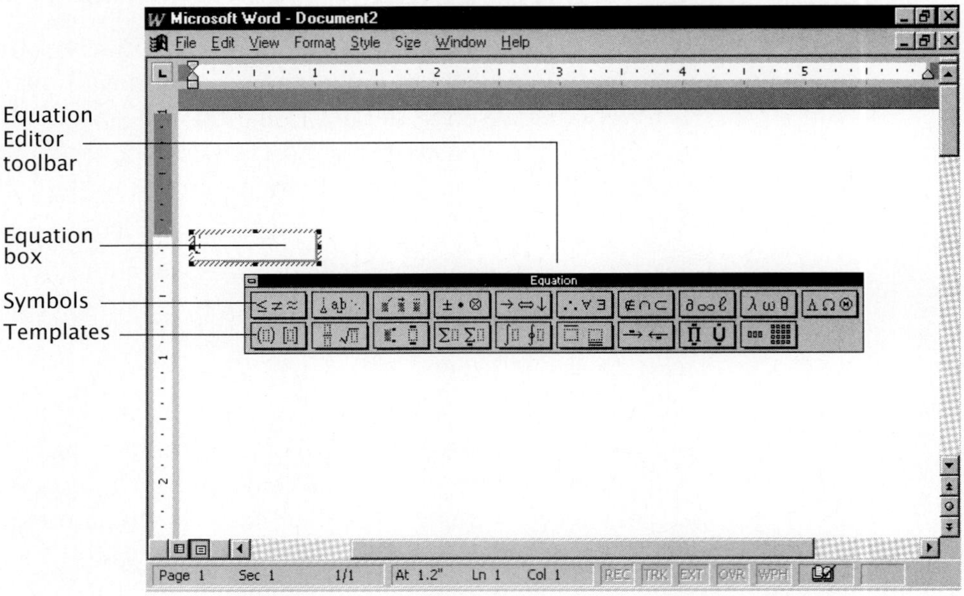

In the upper left corner, the *equation box* is shown. The Equation Menu bar extends across the top of the window, and the *Equation Editor toolbar* is shown in the middle of the screen. To create an equation, you either type the characters or numbers from the keyboard or insert symbols and templates from the Equation Editor toolbar.

Using the Equation Editor Toolbar

The top row of the Equation Editor toolbar contains symbols for you to use in creating equations, and the bottom row contains templates. *Symbols* are items that you merely insert into your equation as single characters. For example, characters from the Greek alphabet, operators such as $\otimes$ or $\bullet$, set operators such as $\subseteq$ or $\notin$, and others can be inserted directly into equations as you create them. *Templates* contain a *slot* that you can fill in with an appropriate number or symbol. For example, the radical symbol ($\sqrt{\ }$) is added to an equation as a template in Word. This template contains a slot that expands in size to accommodate the text, numbers, and symbols that you include under the radical.

The function of each of the symbol buttons on the Equation Editor toolbar is shown next.

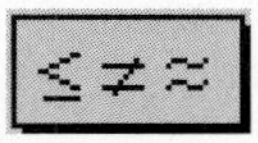

Inserts relational theory symbols, such as $\equiv$, $\approx$, or $\geq$.

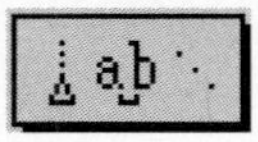

Controls the spacing between text and symbols.

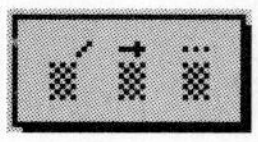
Inserts embellishments in the text and symbols. The types of embellishments available include $\hat{}$, $\ddot{}$, $\leftrightarrow$, and others. Embellishments are added to text or symbols, not separately.

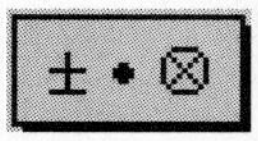
Inserts operators such as $\pm$, $\div$, or $\oplus$.

Inserts arrows pointing to the left, right, up, or down.

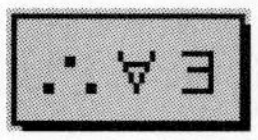
Inserts logical symbols such as $\exists$, $\therefore$, or $\forall$.

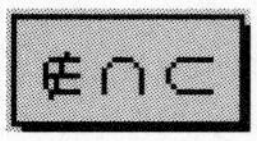
Inserts set theory symbols such as $\supseteq$, $\varnothing$, or $\in$.

Contains miscellaneous symbols such as ∂, ∞, or ∇ for use in Word equations.

Inserts lowercase characters of the Greek alphabet.

Inserts uppercase characters of the Greek alphabet.

The template buttons on the Equation Editor toolbar show the slots associated with the templates as dashed boxes as a part of the graphic for the button. A simple way to create equations in your document may be for you to first create the overall look of the equation by inserting the appropriate templates and then going back and filling in each of the slots by typing or choosing symbols until your equation is complete.

The functions of each of the template buttons on the Equation Editor toolbar are shown next.

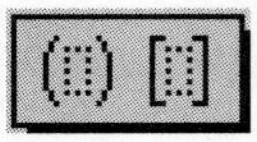
Inserts fence templates such as matrix brackets or parentheses.

Inserts vertical, diagonal, or slash fractions, as well as square-root, nth root, or long-division templates.

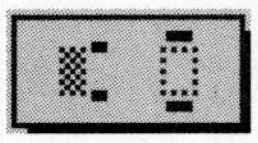
Inserts various superscript and subscript templates.

Inserts various summation templates.

Inserts various integral templates.

Inserts underbar and overbar templates.

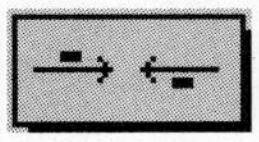
Inserts templates for arrows right, left, or both that have text labels associated with them.

Inserts product and set theory templates.

Inserts matrix templates. You can create matrices that are 1 × 2, 2 × 1, 2 × 2, and so on, or you can create a matrix with customized dimensions.

Entering an Equation

You enter equations by a combination of typing and choosing items from the Equation Editor toolbar. Word will automatically set the spaces between elements of the equation for you. If you want to increase or decrease the spacing, select Spacing from the Format menu. Word will also automatically reduce the size of sub- and superscripts. Many of the symbols found in the Equation Editor can be inserted into your document as special characters without having to resort to the Equation Editor. You will learn about this in Section 7-3 of this module.

If you point to one of the buttons on the Equation Editor toolbar and click the mouse button, a graphical pop-up menu of the items available from that toolbar button will be shown. For example, Figure 5-3 shows the pop-up menu that appears when the lowercase Greek alphabet button is selected from the Equation Editor toolbar. To insert any of these characters into your equation, click the desired letter. You can then continue to make choices from the Equation Editor toolbar to add subsequent symbols to your equation.

Figure 5-3
Lowercase Greek alphabet pop-up menu

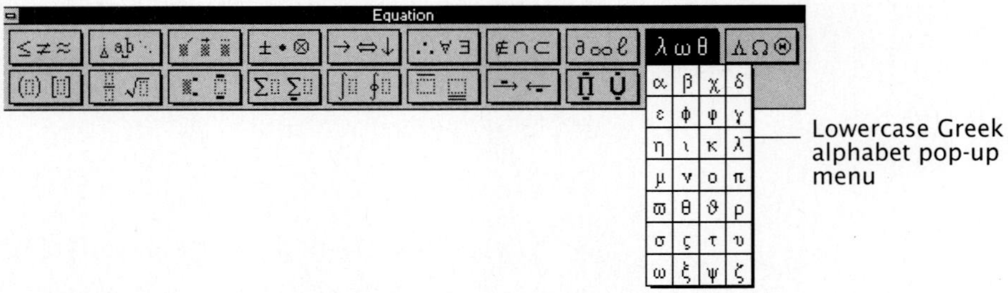

Templates are added to an equation in the same way that characters are added except that after including a template in your equation, you must fill in the slots of the template. For example, Figure 5-4 shows the pop-up menu that is displayed if the fraction and radical template button is selected from the Equation Editor toolbar. If the nth root template is chosen from the pop-up menu, this will be inserted in the equation box as shown in Figure 5-5. Note that this template has two different slots that

Figure 5-4
Fraction and radical pop-up menu

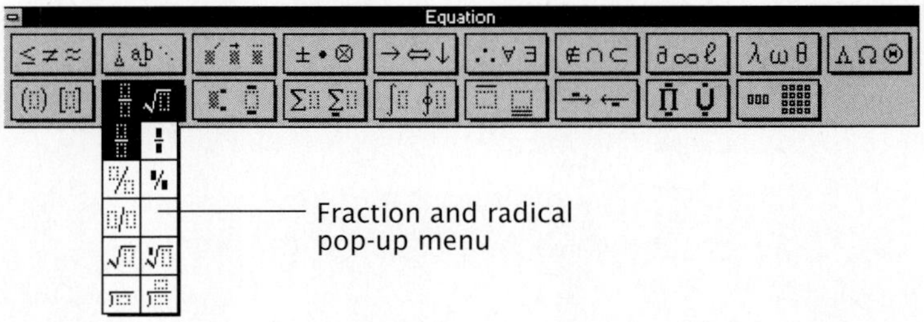

must be filled in—one for the root and one for the argument under the radical. To fill in these slots, place the insertion point at the beginning of the slot, and either type from the keyboard or select a symbol from the Equation Editor toolbar.

Figure 5-5
Equation box with nth root template inserted

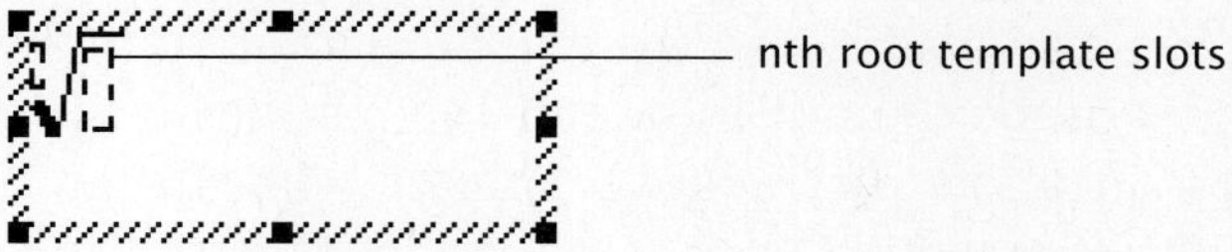

You can also add a template inside either slot. This is known as nesting templates. Figure 5-6 shows the completed equation as it appears in the Equation box.

Figure 5-6
Equation box with completed equation

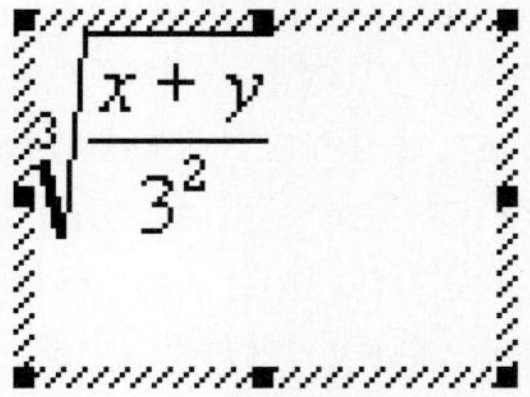

When you finish creating your equation, click anywhere inside the document window to close the Equation Editor and return to editing your document. The equation you have created is placed in the document at the location of the insertion point.

Try It

In this exercise, you create the following equation using the Word Equation Editor:

$$\int x^2 dx = \frac{x^3}{3} \qquad \text{Eq. (1)}$$

- Start a new Word document and open the Equation Editor.
- Create the equation shown by inserting a combination of indefinite integral, superscript, and large vertical fraction templates.
- Close the Equation Editor to return to your Word document. Save the document under the filename PROPOSE1.DOC.

Editing Equations

You may need to move your equation after creating it. Before you move an equation, you need to select it by clicking it. The equation will be surrounded by a set of square sizing *handles* as illustrated in Figure 5-7. Equations are regarded as single characters within a Word document. Once an equation is selected, you can cut, copy, or paste it anywhere in your document. You can also set up the equation to be in a paragraph by itself and then use any of the justification or margin commands to control the placement of the equation within the document.

Figure 5-7
Equation box selected

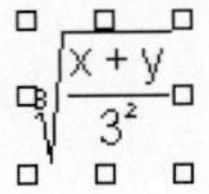

You can also resize the equation after you have created it. To resize an equation box, point to one of the handles on its border and drag it to its new location. The mouse pointer will change appearance to a double-pointed arrow when you are resizing your equation. To resize only one graphics border at a time, drag one of the middle handles. To resize in two directions at the same time, drag one of the corner handles. This will actually change the size of your equation within your document. All the characters within the equation will be scaled by the same amount so that the proportionality between the characters will remain the same after resizing.

Once you have an equation in your document, you may wish to edit or delete it. To delete an equation, select it and press DEL or BACKSPACE. To edit an equation, double-click it. The Equation Editor appears, displaying the selected equation. At this point, you can edit your equation much as you would your document (selecting parts of the equation, cutting, pasting, copying, changing fonts), using the options in the Equation Editor menus and toolbar. To change the font of the characters in the equation, choose Other from the Style menu and apply the font of your choosing. To change the text size, choose Other from the Size menu and specify a new size for the characters. To delete a particular template from an equation, you must first delete its contents.

Try It

In this exercise, you edit the equation that you created in the previous Try It! exercise.

- Open PROPOSE1.DOC, which you created in the last Try It! exercise, if it is not already open.
- Select the equation within your document, and then resize and move the equation to a new location within the document.
- Double-click the equation to return to the Equation Editor. Delete the contents of the indefinite integral, and then delete the integral template. Insert a new template that is an integral with overscript and underscript limits.
- Fill in the slots in this template so that the lower integration limit is 1, the upper limit is 2, and the integrand is x^2dx.
- Edit the right side of the equation so that it reads $\frac{2^3}{3} - \frac{1^3}{3} = \frac{7}{3}$.
- Change the equation font to Arrus BT and the font size to 16 point. If your lab setting does not support this font, use Courier.
- Return to your document. Your equation should now look like the one shown here. Save the document as PROPOSE2.DOC.

$$\int_1^2 x^2 dx = \frac{2^3}{3} - \frac{1^3}{3} = \frac{7}{3}$$

Using Equations in Documents

In engineering documentation, equations are usually set apart from the text by adding extra space above and below. Immediately following the equation, you should explain the meaning of all the terms in the equation, as in the example shown in Figure 5-8.

Figure 5-8
Correct use of the equation within a segment of text

For a simple tension test, the relationship between the applied stress and the measured strain is given in the following:

$$\sigma = E\varepsilon \qquad \text{Eq. (1)}$$

where σ is the stress in units of psi, E is the modulus of elasticity in units of psi, and ε is the strain in units of in/in.

You must also number equations in the order in which they appear in the document. Note that the equation shown in Figure 5-8 includes the right-justified number "Eq. (1)." Such numbers enable you to refer to equations easily. For example, you could say something like "Rearranging Eq. (1) and substituting the result into Eq. (3) yields the following" in your document.

You can insert equation numbers in Word documents as captions. Captions are labels or titles you add to either graphics or equations. You can have Word automatically include captions by selecting Caption from the Insert menu. From the Caption dialog box shown in Figure 5-9, you can select a label of either Equation, Figure, or Table, and select a caption position of either above or below the selected item. You can also create a customized label by selecting New Label from this dialog box. The main advantage in using Word captions is that they are automatically numbered for you and will be updated if equations are added, deleted, or rearranged in your document.

Figure 5-9
Caption dialog box

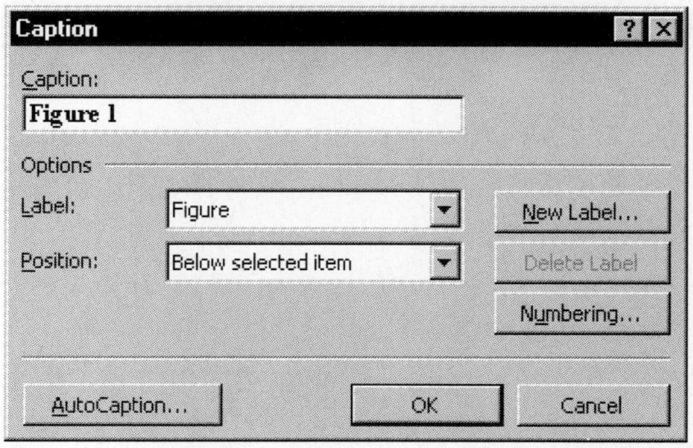

Unfortunately, there is no easy way to insert right-justified captions on the same line as the equation, which is normally how equation numbers are displayed in technical documents. To add right-justified equation captions, set a right tab stop at the right page margin, and use (TAB) to add the equation caption at the right margin. To center the equation on the page, select it and then drag it to its new location in the document.

Although this method works, it is tedious. Another problem with this method is that your equation numbers will not automatically be updated as you edit your document.

Try It In this exercise, you add a caption to the equation you created in the previous exercises, and you add some text to your document.

- Open PROPOSE2.DOC if it is not already open. Select your equation, create a new label for a caption so that it reads EQ. 1, and add a caption below the equation.
- Place the insertion point above the equation in your document, and type the following sentence:

  `The following represents the area under the curve for the function x`^{2}`. The limits of the integration are 1 and 2.`

  (Note that you have to use the superscript feature in the Font dialog box to accomplish this; refer to Chapter 3 if you need to refresh your memory on how to do so.)
- Below the equation, type the following text:

  `where x represents any variable.`
- Print the document.

SUMMARY

In this chapter, you learned about writing engineering proposals. Proposals typically describe an investigation to be conducted during the third stage of the design process. In this stage, you investigate alternative solutions to the problem, evaluating their feasibility and their compliance with the design criteria. Most engineering proposals include equations to describe the results of previous investigations. This chapter explained how to create, edit, and number equations within a Word document.

Key Words

caption
equation box
Equation Editor toolbar
handles
proposal
slots
symbols
templates

Exercises

1. Create the following equation in Word:

$$i = \frac{75}{5 + R}$$

2. Create the following equation in Word:

$$y = \int_{0}^{t} v(t)\, dt$$

3. Create the following equation in Word:

$$Flux = -D\frac{dc}{dx}$$

4. The velocity of a particle is described by these three relationships:

$v = 10t^2$ m/s	$0 < t < 1$ sec
$v = 1000 - 5t$ m/s	$1 < t < 2$ sec
$v = 45t + 2(t - 20)^2$ m/s	$2 < t < 3$ sec

In Word, write the equation that could be used to solve for the total distance traveled by the particle. (Recall that the distance is the area under the velocity curve.) Then write the equation that could be used to solve for the acceleration of the particle. (Recall that acceleration is the derivative of the velocity curve.)

5. Write the following equation in Word:

$$f(Q) = 0 = \sqrt{1 + \frac{1}{aQ^2}} + \frac{1}{2Q} - 1.125$$

6. Create the document shown. Change the equation font to 14-point, bold, Arial font. (If your printer cannot print this, change the font to something else.) Italicize the variables as they appear as text within the document. Make sure to number the equations.

An electric circuit is forcing a time varying current through an induction coil of 2×10^{-3} Henries. The voltage that will be developed is

$$v = L\frac{di}{dt} \qquad \text{Eq. (1)}$$

where *v* is the voltage measured in volts, *L* is the inductance of the coil measured in Henries, *i* is the current measured in amps, and *t* is time. The power passed to and from the coil is:

$$p = vi \qquad \text{Eq. (2)}$$

where *p* is the power measured in watts. The energy passed to and from the coil is given by

$$w = \frac{1}{2}Li^2 \qquad \text{Eq. (3)}$$

with energy, *w*, measured in joules.

7. Write a short proposal outlining the procedure you would use to determine the velocity and acceleration of a falling particle. Use at least two equations in your proposal.

8. In setting safe curve speeds, traffic engineers use a special type of level. They drive around the curve and determine which speed causes a tilt of 10 degrees. Using Word, write a proposal outlining the procedure you would use to determine the safe speed for a section of highway that has several curves. Use at least two equations or mathematical expressions in your proposal.

6 Writing Technical Papers

Telecommunications

The telegraph was the most important form of rapid communication in the 19th century. The telegraph permitted people to convey short messages quickly from one point to another. However, telegraph lines were subject to frequent breakdowns, and signals could be conveyed only a limited distance and had to be relayed from one point to another along the system. Currently, fiber-optic networks and microprocessors have been combined to form a new and worldwide network. Information can be communicated efficiently and almost instantaneously across the network, making worldwide communication nearly effortless. In the future, this worldwide network will change the ways that we gather information and communicate with others even more significantly. In this chapter we explore the fourth stage in the design process—choosing the best solution—and the way that engineers communicate with others through technical papers during this stage.

INTRODUCTION

After investigating alternative solutions, the engineer must select the one solution that best satisfies the design parameters. In making this decision, the engineer must weigh each of the design parameters against one another. The solution may be the one option that best meets all the design criteria; the one that meets the most important design criteria optimally and does an adequate job with the remainder of the criteria; or the least-expensive solution, with marginal adherence to the design criteria.

This chapter discusses the fourth stage in the design process. At this stage, each of the possible solutions is weighed, and the best solution is determined. You will learn about technical papers, the type of documentation that you are most likely to create at this stage, and about making oral presentations.

6-1 CHOOSING A SOLUTION

After closely examining all the alternative solutions, you must choose the best solution and solve the problem. To come up with a workable solution to solve the problem at hand, you need to apply the knowledge that you have gained from your formal education and from practical experience.

For example, suppose you are an electronics engineer in charge of designing the circuitry for a computer power supply. At some point in the design process, you may have to draw on your engineering knowledge to design a voltage divider. A voltage divider is used when an electronic component has an allowable voltage range that differs from the 120 volts available from a standard outlet. The schematic diagram for a voltage divider is shown in Figure 6-1.

Figure 6-1
Electrical circuit

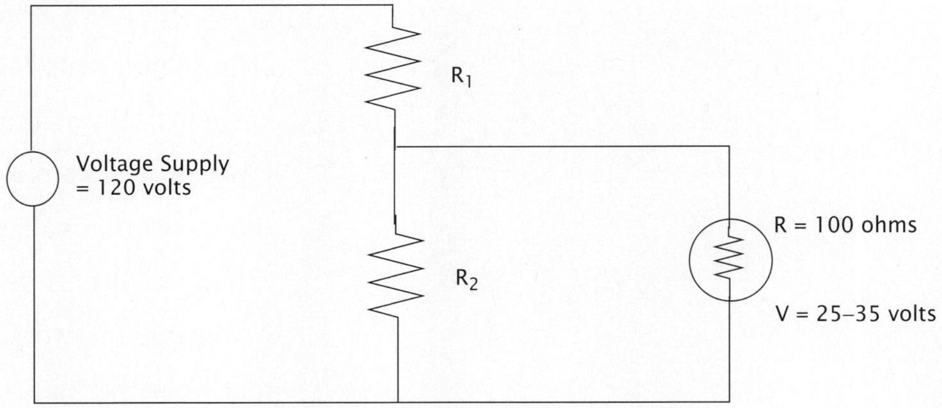

In this problem, from the principles taught in physics and introductory circuits courses, the following is the governing equation:

$$R_{EQ} = \frac{R R_2}{R + R_2} = \frac{100 R_2}{100 + R_2} \qquad \text{Eq. (6-1)}$$

where R_{EQ} is the equivalent resistance, and R (=100 ohms) is the resistance of the component. The allowable range for the component is

25 to 35 volts. The following equation can be used to determine resistances R_1 and R_2:

$$V = 25 - 35 \text{ volts} = 120 \frac{R_{EQ}}{R_1 + R_{EQ}} \quad \text{Eq. (6-2)}$$

As an electronics engineer, you would use these two relationships in a trial-and-error solution to determine the two unknown resistances. Some of the design restraints might be that you want to minimize the wasted power in the circuit (this means that the two resistances should be as large as possible) and that you want to maintain a voltage of between 25 and 35 volts within the component.

You usually encounter calculations like these in the fourth stage of the design process. Depending on your field, the variable names and formulas will change, and the problem itself may be more or less difficult. What remains the same is that during the fourth phase of the project is when the actual design is accomplished. A typical design consists of drawings, calculation sheets, and a written document in the form of a technical paper or report. This chapter describes how to write technical papers and make technical presentations. You will learn how to create and include figures within a Word document because technical papers commonly contain figures, as well as equations and tables.

6-2 FORMAT OF TECHNICAL PAPERS

When writing *technical papers*, you must write in a clear, concise style. Generally, the passive voice is preferred, and you shouldn't use the first person. However, in some cases, you can use both the active voice and the first person. Again, it is best if you know your audience. Technical papers are a crucial component of engineering design documentation because they give the overall view as well as some of the details about how a design was accomplished. Here is the format of a typical technical paper:

Title and author — Frequently, the title and author are included on a separate title page, but this is not always necessary.

Table of contents — If the paper is fairly long, include a table of contents so that readers can find topics quickly and easily.

List of figures — Technical papers often include several figures. It is helpful to supply the figure names and page numbers in a separate list.

Abstract or summary — A well-written technical paper usually begins with a short, introductory abstract or summary mentioning the major results or findings that are discussed in detail in the paper itself.

Introduction	This section includes any pertinent background material and the objectives of the design process. The introduction should also clearly state the problem being addressed.
Main body	The bulk of the paper goes in this section. Here you outline the design process as it was applied to the specific problem. You also discuss which alternative solutions were considered and why the final design was selected. The main body of the paper is usually divided into separate sections and subsections.
Discussion and conclusions	In this section, you discuss some of the specific features of the design and any problems that you encountered in the design process. Your conclusions often contain recommendations for future projects or future studies.
References	If your paper cited books or journal articles, you need to include those references here. In citing and listing references in technical papers, follow the procedures outlined in Chapter 5.
Appendixes	The appendixes contain any pertinent data or intermediate calculations. It is especially important to include this material so that anyone trying to retrace your steps will have all the necessary information.

6-3 CREATING A TECHNICAL PAPER IN WORD

Technical papers typically incorporate all the features covered so far in this module, including customized paragraphs and characters, tables of data, and equations. In addition, technical papers almost always contain figures that illustrate a point within the text. You can use Word features to include figures as a part of the document itself—you can cut and paste electronically instead of having to physically cut and paste figures into the document.

The remainder of this chapter describes how to incorporate figures within a document. You can create simple line drawings in Word itself. However, it is probably better to create complex figures with an external drawing package such as AutoCAD® and then import the figures into your Word document. You learn how to import graphics from external programs later in this chapter.

Creating Simple Line Drawings

To create simple diagrams in Word, choose Object from the Insert menu. The Object dialog box appears (see Figure 5-1). Choose Microsoft Word Picture from this dialog box. The screen appearance will change so that it is similar to that shown in Figure 6-2. Note that the actual drawing area (inside the picture boundary) is relatively small. You may wish to resize it by dragging one of its corner points.

Figure 6-2
Word drawing screen

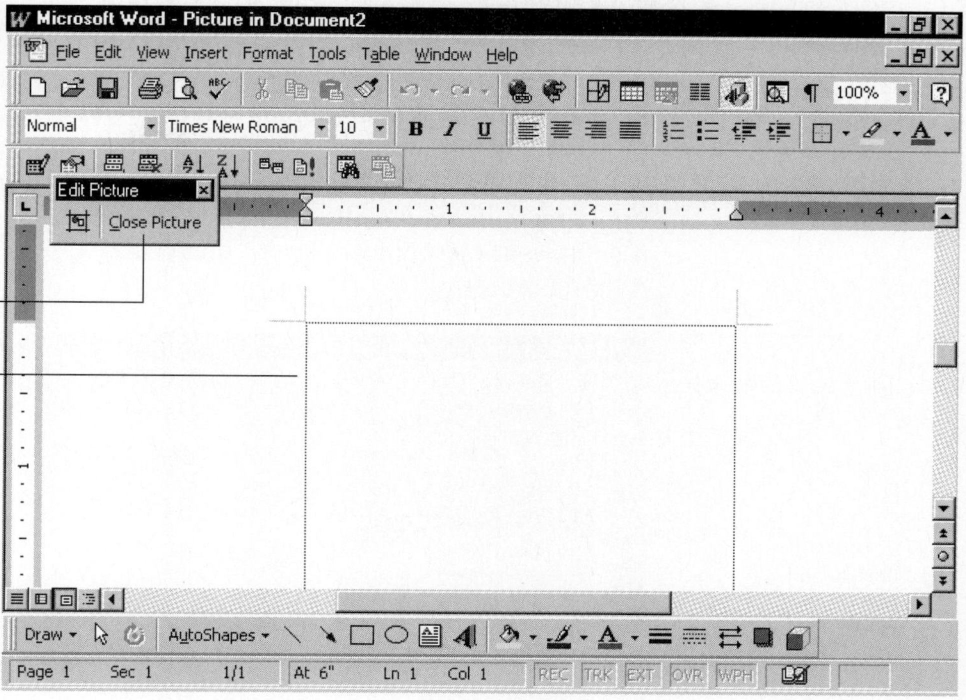

The picture boundary is visible in the main part of the drawing editor screen. You draw your figure inside this boundary. When you are finished, everything within the boundary will be inserted as one entity within your document. You can then move this picture within your document, cutting and pasting it as necessary. Along the bottom of the screen is the *Drawing toolbar.* With the options found on this toolbar, you can create and edit your Word drawings.

Drawing Objects

Before you begin to draw, you must select the appropriate button from the toolbar. You use the mouse to draw in Word. The status bar of the document window identifies how each button on the toolbar is used.

The Drawing toolbar options that you use to draw objects are described here. The remainder of the Drawing toolbar buttons will be described later in this chapter.

Used to draw lines by clicking the first endpoint and dragging to the second endpoint.

Used to draw arrows by clicking the first endpoint and dragging to the second endpoint (similar to the way you draw lines).

Used to draw rectangles by clicking the first corner point and dragging to the corner point located across the diagonal from the first point.

Used to draw ovals by clicking the first point and dragging to the second point. You can also draw a circle with this button by holding down the SHFT key while dragging.

In addition to these simple drawing tools, you are able to create several different predefined shapes in your document by selecting any one of the Autoshapes available in Word 97. To select an Autoshape, point to that option on the Drawing Toolbar and click. A pop-up menu appears with six different options: Lines, Basic Shapes, Block Arrows, Flowcharts, Stars and Banners, and Callouts. If you select any of these options, a graphical pop-up menu will appear displaying the types of shapes you can draw. Figure 6-3 shows the graphical pop-up menu that appears when the Basic Shapes option is selected.

Figure 6-3
Basic Shapes Graphical Pop-up menu

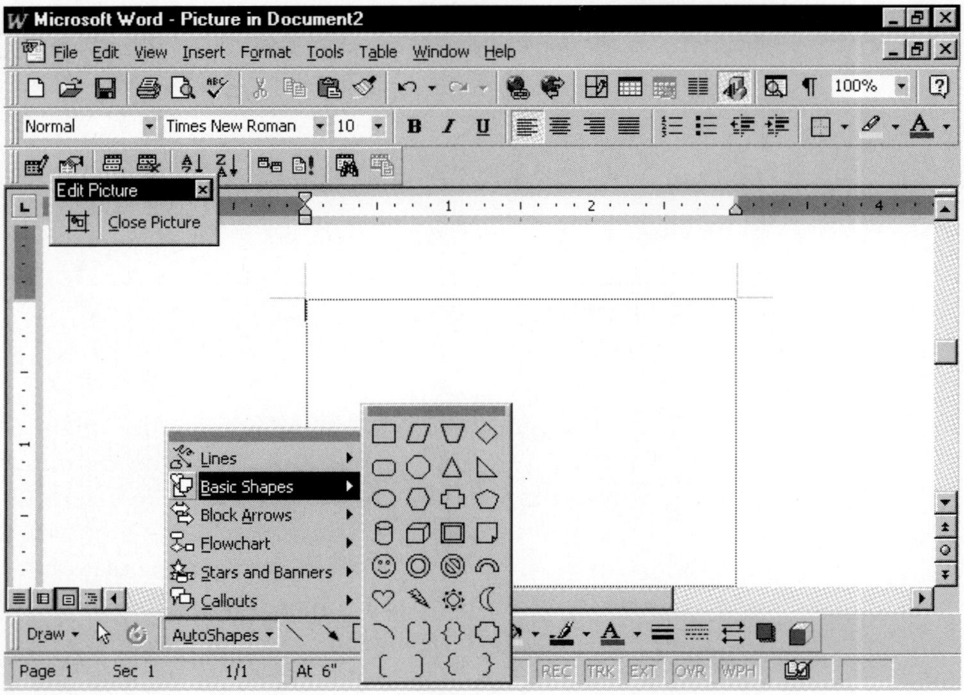

When you are drawing Word objects, you will see a feature known as “rubber-banding.” For example, after selecting the first endpoint of a line, the “line” will follow the mouse pointer as you drag to its other endpoint. This feature enables you to see exactly what the line will look like before you actually draw it.

Using Snap to Grid

Most of the time when you are making Word drawings, it is helpful to turn on the *Snap to Grid* feature. The Word drawing editor includes an invisible *grid* within the picture frame. When Snap to Grid is turned on, the only points that you are able to select in the picture frame are the grid points. Thus, you are better able to control the spacing and location of the points that define the geometry. To turn on this feature, select Grid from the Draw menu found on the Drawing toolbar.

A dialog box similar to the one shown in Figure 6-4 appears. You can use the options in this dialog box to control the *grid spacing* in either the vertical or horizontal directions and to control the grid origin. You can also turn the Snap to Grid feature on or off by alternately selecting or clearing the Snap to Grid check box in this dialog box.

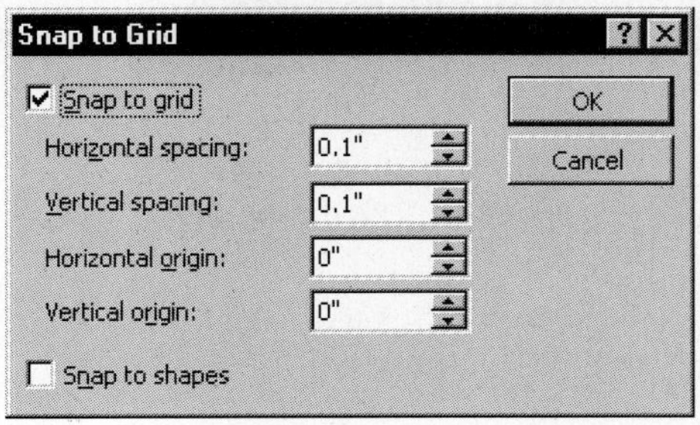

Figure 6-4
Snap to Grid dialog box

Editing Objects

When making drawings, it is sometimes necessary to select existing objects for editing. To select an object, point to it and click. Once an object is selected, handles similar to those that appear when editing equation boxes will surround the object.

You can now move, resize, cut, copy, or paste the object anywhere on the drawing screen. To move the object, point to it and drag it to its new location. To resize an object, drag one of the handles defining it. You can also choose Cut, Copy, or Paste from the Edit menu to perform these operations on the selected object.

Most of the options for editing objects are found in the Draw pop-up menu (accessed from the Drawing toolbar). Some of the most frequently used options are described here:

Group. This option groups multiple selected objects so that they are a single entity in the drawing. To select more than one object for grouping, hold down SHFT as you select the objects. After objects have been grouped, they can be moved, resized, copied, or pasted as a single entity.

Ungroup. This ungroups objects that have previously been grouped together.

Order. This option has several available operations: (1) Bring the selected object to the front of the drawing. Other objects will be covered up by this object as appropriate. (2) Send the selected object to the back of the drawing. Parts of the selected object may be hidden from view by other objects after this is selected. (3) Bring the object forward. (4) Send the object backward. (5) Bring the selected object in front of a text box. (6) Send the selected object to the back of the text box.

Nudge and Align or Distribute. These options allow you to move selected objects up, down, left or right or to align them with other objects in the drawing,

Rotate or Flip. You can choose to flip the selected object from left to right or from top to bottom. You can also choose to rotate the object 90 degrees to the right (clockwise) or 90 degrees to the left (counterclockwise). Another option allows you to freely rotate an object by dragging one of its control points on the drawing.

Edit points. This allows you to reshape any shapes on your drawing. When this is selected, the control points, or points that define the shape, appear as handles that can be dragged to new locations to change the shape of the object.

Try It In this exercise, you use Word to create the diagram of an electrical circuit shown here.

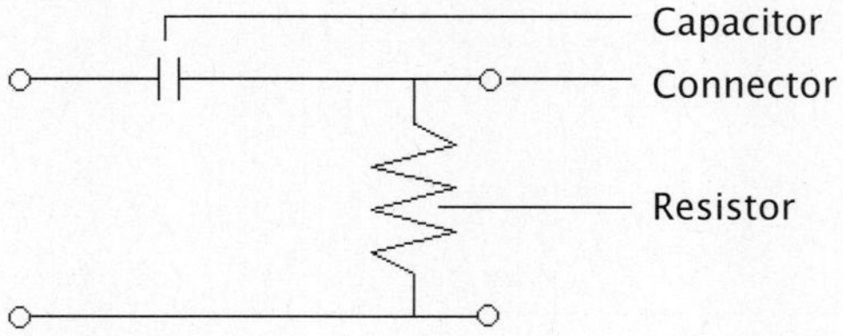

- Start Word, and start up the drawing editor.
- Turn on the Snap to Grid feature.
- Start by drawing the resistor shown on the right side of the diagram. It is probably best to draw this as a free-form polygon where points on the polygon are defined by clicking rather than dragging. If you make mistakes when creating the resistor, keep going. You can reshape the polygon to make things line up better after it is already drawn.
- Now create five single lines—three horizontal lines for the circuit and two vertical lines for the capacitor.
- Next draw one small circle for the connecting points as seen in the figure. You will copy this circle in the next step. To draw the circle, you probably want to turn the Snap to Grid feature off. You may also want to draw the circle anywhere on the drawing and then move it to the desired location.
- Copy and paste this circle to its three other locations on the drawing. Note that when you copy and paste the circle onto your drawing, the copy may initially appear on top of the original. You can then drag this copy to the correct location on the drawing.
- Save your drawing in a file called `CIRCUIT1.DOC`.

Changing Object Attributes

Attributes define the way that an object appears in the drawing. The three attributes you change most often are the line style, the fill style, and the colors. Word97 also has attribute settings for dash style, font color, shadows, and 3-D effects. You must first select an object in order to change its attributes.

The Drawing toolbar buttons that are used to change object attributes are outlined next.

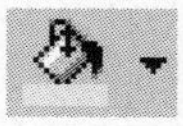

Sets the fill pattern and color for objects. Click the arrow on the right to obtain a graphical pop-up menu of the available colors and patterns. Objects do not have to be completely enclosed in order to apply fill patterns to them. Some of the fill styles available are differing shades of gray.

Sets the line color for objects. Click the arrow on the right to obtain a graphical pop-up menu of the available colors and patterns.

Sets the color for text you have created in your drawing. Click the arrow on the right to obtain a graphical pop-up menu of the available colors.

Sets the line style for objects. You can change the line thickness or change a line so that it is a double line. You can also add arrowheads to the beginning, the end, or both ends of a line or curve by selecting the More Lines option from the graphical pop-up menu. You can also change the line style so that it appears as a dashed line with either long or short dashes. The changes you specify are applied to single lines as well as to the lines defining polygons, circles, ellipses, and so on.

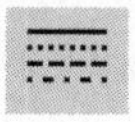
Sets the dash style for lines or curves. Your choices include short dashes, long dashes, dots, crosses, or a combination of long and short dashes.

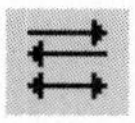
Sets the arrow style for lines. Your choices include solid, hollow, circles, both ends, or just one end of the line.

Sets the shadow effect for a given object.

Sets the 3-D effect for a given object.

Adding Text and Labels to Drawings

When you create figures for technical papers, you typically need to add *notes* or *labels* (also named *callouts*) to the drawings. Both notes and labels contain text, but labels also include an arrow or line pointing to some feature of the drawing.

When adding text to a figure, you should first turn off the Snap to Grid feature so that you are free to choose any point on the screen. Add text to your drawing by selecting the Text Box button on the Drawing toolbar as shown in Figure 6-5. When you select the Text Box button, you include a text box in the drawing by clicking the first corner point on the box and dragging to the corner point located across the diagonal from the first point. Text is often added to drawings to label parts of the figure or to make the figure more understandable.

Figure 6-5
Text Box button on Drawing toolbar

Callouts are added to your drawing by selecting the desired type from the graphical pop-up menu that appears when Callouts is selected from the AutoShapes menu. A callout includes an arrow pointing to a feature of the drawing and a text box. Callouts are created by clicking the location of the arrowhead and dragging to the point where the text box will be located.

When you are creating text boxes and callouts on your drawing, the borders of the text box may overlap other objects in your drawing. You can change the size of the text boxes as needed by first selecting the text box and then dragging a handle on the box to resize it. Word will automatically create a text box border in your drawing with a thin solid line around the text. Many times you will not want this border to appear in your drawing. To change this, first select the text box you wish to edit. Select the Line Style button on the Drawing toolbar, and from the graphical pop-up menu that appears, choose More Lines. A dialog box will appear. From this dialog box, choose the down arrow next to the Color list box, and then specify No Line as the Line style for this text box.

Exiting the Drawing Editor

As you are creating your drawing, you may end up drawing in the area outside the picture frame. When you wish to insert the figure into your document, you should first resize the picture boundary so that all the objects and text boxes fit within the boundary. To automatically resize the picture boundary to fit all objects and text boxes, use the Resize Picture Boundary button in the Edit Picture dialog box. This dialog box is shown in Figure 6-6.

Figure 6-6
Picture dialog box

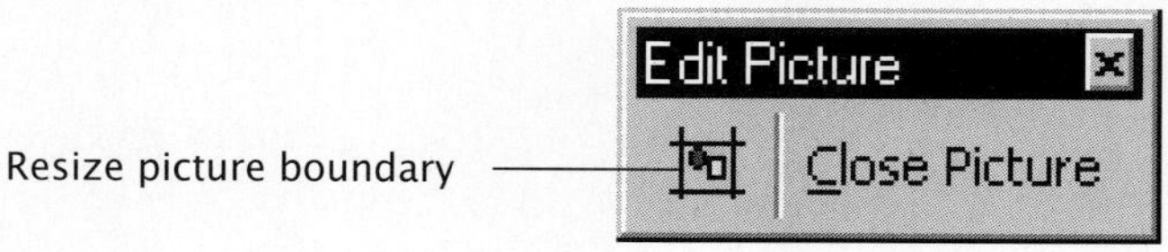

When you are finished creating and editing your drawing, you return to your document by selecting Close and Return to (Document name) from the File menu or by selecting the Close Picture button on the Picture dialog box (see Figure 6-6). Word then inserts the picture frame within your document at the location of the insertion point. You will not be able to see your drawing unless you are in Page Layout view.

If you want to change any of the individual objects within a figure after the figure is inserted in your document, point to it and double-click. Control will automatically revert to the drawing editor, and the figure will be ready for editing. After you have exited the drawing editor, the Picture toolbar will be displayed across the top of the screen.

Adding Figure Captions

A technical paper's figures are numbered consecutively and given descriptive titles. For example, your document might include a figure labeled "Figure 1. Schematic of Circuit A." In Word, you add figure titles as *captions*, just as you did when numbering equations. (If necessary, refer to Chapter 5 for the details on creating captions.) Unlike equation numbers, however, figure captions go beneath the figure. After selecting the figure, choose Caption from the Insert menu to include a caption below the selected picture frame.

Try It

In this exercise, you add text and labels to the drawing you created previously. When you are finished, your drawing should look like the one shown here.

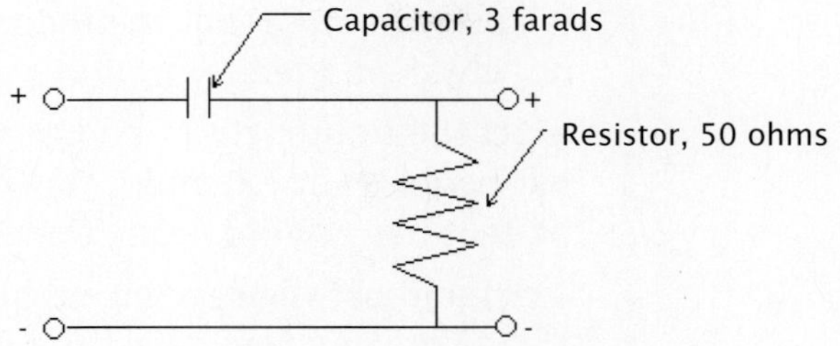

- Open CIRCUIT1.DOC if it is not already open. Turn off Snap to Grid.
- Add figure callouts for the resistor and capacitor. Change the line style of the callout to include arrowheads, and resize the text boxes as necessary.
- Add text to your figure to label + and – connections. Remove the text box borders from these. Your drawing should be similar to the drawing shown.
- Resize the picture boundary so that the entire drawing is inside the frame.
- Close the picture, and return to your document.
- Add the caption "Figure 1. Partial Electrical Circuit." Save the document as CIRCUIT2.DOC.

Moving and Resizing Figures

You can move and resize figures within your Word document much as you moved and resized equations. If you click the figure, handles appear. You can drag these handles to resize the figure; you can also cut, copy, and paste the figure as well as center-justify the paragraph that contains the figure. Like resizing equations, when you resize a figure, the size of the figure increases or decreases as you drag the graphics box handles. The caption moves along with the figure.

Try It

In this exercise, you edit your drawing and move it within your document text.

- Open CIRCUIT2.DOC if it is not already open. Select the figure, and resize the graphics box so that it stretches across the entire width of the page. (Note that the size of the drawing within the figure increases by a proportionate amount.)
- Type the following paragraph above the figure:

  ```
  In the analysis of an electrical circuit, the transfer
  function is often of interest. This function can be used
  to determine the effect of the circuit on a sine function
  in terms of magnitude changes and phase changes. For exam-
  ```

`ple, consider the following circuit, which contains a capacitor and a resistor. Kirchhoff's voltage law can be used to derive the transfer function for this system.`

As you type, note that the figure automatically moves down to accommodate the added text.

- Select the figure, cut it, and then paste it after the end of the sentence that begins "For example, consider... ." Note that the last sentence of the text has automatically flowed around the graphics box.
- Select and cut the sentence that begins "For example, consider... ." (Make sure you also select the picture.) Paste the sentence after the first sentence in the paragraph. Note that the sentence and the figure move, and the remainder of the text flows around the graphic.
- Save your document as CIRCUIT3.DOC for use in a later exercise. Print the document.

Importing Graphics from Other Sources

Sometimes you will want to bring graphics files created in other software applications into your Word document. Word allows you to import graphics files of many different formats, including Encapsulated PostScript files, Windows Bitmaps, Macintosh PICT files, and DrawPerfect graphics files.

To import a graphics file into your document, place the insertion point in the desired location and choose Picture from the Insert menu, and then choose From File from the subsequent menu. A dialog box similar to the one for opening or saving files appears, as shown in Figure 6-7.

You can specify the path to the file and the file type of the required graphics file the same way you specify a path when opening a document. Word lists all the files of the specified type found in the specified path. To change the file type, select the Files of Type drop-down list box, scroll through the list, and select the desired file type. When you are done, choose OK to bring the graphics file into your document.

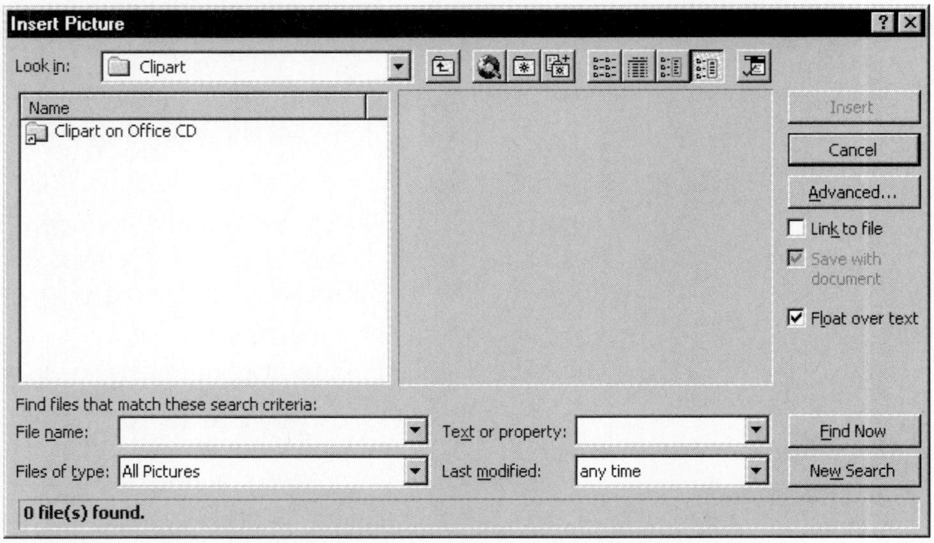

Figure 6-7
Insert Picture dialog box

Try It In this exercise, you start a new Word document into which you import an externally created graphics file. You use a file originally created in AutoCAD and converted to a TIFF file named CIRCUIT.TIF. Ask your instructor how to find this file in your laboratory setting.

- Start a new Word document.
- Choose Picture and then From File from the Insert menu.
- In the dialog box that appears, list all files of type All Pictures. Specify the path to the file CIRCUIT.TIF, and choose OK.
- In a moment, the figure should appear in your document. Add your name to the document, and obtain a printout to turn in to your instructor.

6-4 MAKING ORAL TECHNICAL PRESENTATIONS

As an engineer, you will often have to make formal oral presentations describing your design solution. The audience for your presentation could include your immediate supervisor, your colleagues, your clients, citizen groups, or people employed in governmental agencies. Be sure to tailor your presentation to fit your audience. For example, if your audience is a group of citizens concerned about the environmental ramifications of a proposed project, try to make your presentation's content less technical.

Technical presentations almost always involve some sort of *visual aids*—either slides, *overheads* (also called transparencies or viewgraphs), or flip charts. You can also purchase presentation graphics software that lets you create slides on the computer and then project them directly onto a screen with a special type of overhead projector. This section focuses on using Word to create overhead transparencies and provides you with some general rules about public speaking.

You can first compose overheads with Word and then copy them onto transparencies. Sometimes you can print overheads directly onto transparencies. Check with your computer support specialist to determine the best method in your computer laboratory. You should use a font size of 18 or 24 points and include only a few statements per page. Novices often make the mistake of supplying too much information on each overhead. If you do this, the audience spends more time trying to read everything on the overhead than paying attention to what you have to say. You should also enlarge equations and figures so that they are clearly visible on overheads.

Your first overhead should include the title and author of the paper to be presented. The second overhead should contain a general outline of the presentation so that the audience knows what you intend to speak about and can follow your line of reasoning. Subsequent overheads should present the most important aspects of the paper in an outline format. The overheads should not contain all the detail given in the paper. It is your job to supply these additional details orally. If you include an enlarged figure as a part of a visual aid, you will probably need a pointer so that you can point to the aspects of the drawing you are speaking about. At the end of your presentation, include an overhead that summarizes the main points of the talk and possibly an overhead that contains your recommendations for further work or research.

The first time you make a technical presentation, you are likely to be nervous. But your confidence will increase each time you speak in front of a group, and eventually you will become more relaxed. You may even find that you enjoy speaking in public.

Before you deliver your presentation, practice a few times either by yourself or in front of a small group of friends or colleagues. This is especially important if you are giving an oral presentation for the first time. Review each overhead, and decide what you are going to say at that point in the presentation. You may find it helpful to write out a script beforehand. If you have a limited amount of time for your presentation, check that you can cover your material in the allotted time. Also allow enough time at the end of the presentation for questions from the audience.

The day of the presentation, make sure to dress professionally. During the presentation, speak in a clear and distinct voice. Try to speak in complete sentences, and maintain eye contact with your audience. Whenever possible, avoid speaking verbatim from your notes. With practice, you should be able to make a smooth, professional presentation. When you are finished, invite the audience to ask questions.

Creating Overheads

When creating overheads, you will often need to copy parts of one Word document to another document. In Word, you can open more than one document at a time. You can then cut, copy, and paste material between these documents just as you cut, copy, and paste within a single document. To switch between documents, choose the name of the desired document from the Window menu.

For example, suppose you want to copy a figure from document A to document B. With document A visible, select the figure (the handles will appear). Then choose Copy from the Edit menu. Select the filename corresponding to document B from the Window menu to display that document. Place the insertion point where you want to put the figure, and choose Paste from the Edit menu to insert the figure into the second document. You can also copy and paste blocks of text or equations in the same manner. When making overheads, you will probably want to enlarge the text, figures, and equations in the new document so that they are easy to see during your presentation.

Try It In this exercise, you create a separate overhead that contains some text and the figure you created earlier.

- Open `CIRCUIT3.DOC`.
- Start a new document.
- Type the following text:

    `Simple Electrical Circuit:`

- Change the font size of this text to 24 points, and change the font style to bold. Center the text between the left and right margins.

- Display the document CIRCUIT3.DOC by selecting its name from the Window menu. Select the figure from this document, and then copy and paste it into your new document. Position the figure below the text you just entered in the new document.
- Print this document.

6-5 USING SHORTCUTS

Opens the Drawing toolbar.

SUMMARY

In this chapter, you learned how to write technical papers during the fourth stage of the design process—the stage at which the solution is chosen and the design is accomplished. You then learned how to add graphics to documents by using Word. Finally, you learned about making technical presentations and using Word to create overheads for presentations.

Key Words

attributes
callouts
captions
Drawing toolbar
grid
grid spacing
import
labels
notes
overheads
Snap to Grid
technical papers
technical presentations
visual aids

Exercises

1. Using Word, create the following figure:

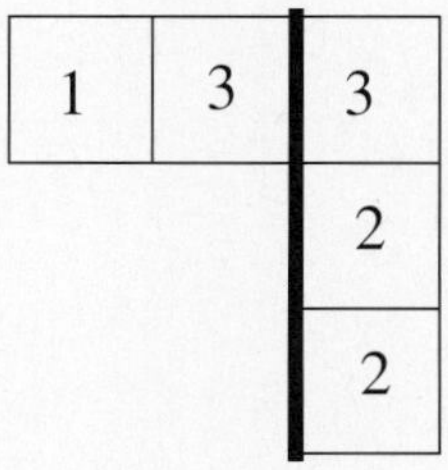

Figure A. Building coded plan

2. Using Word, re-create Figure 6-1.
3. Using Word, create the following figure:

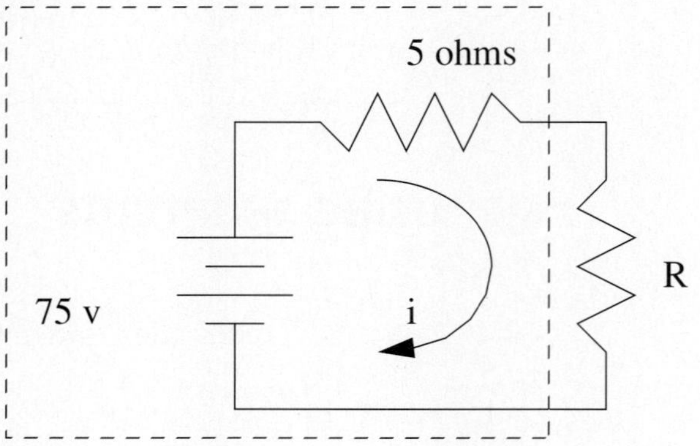

Figure B. Model of a battery

4. Using Word, create the following figure:

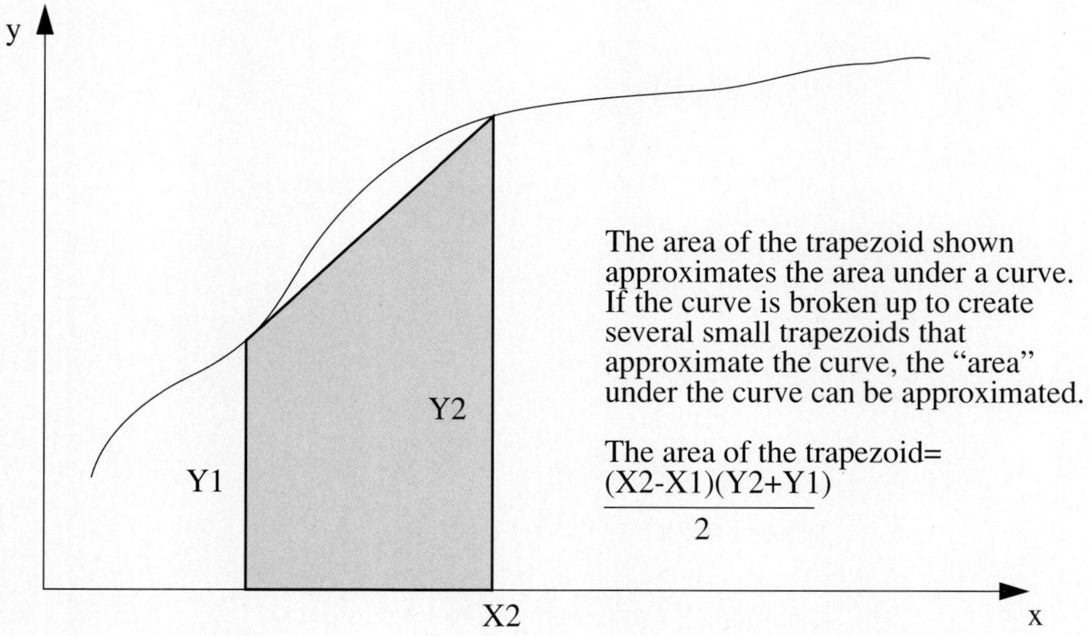

Figure C. Integration by the trapezoidal rule

5. Using Word, create the following figure:

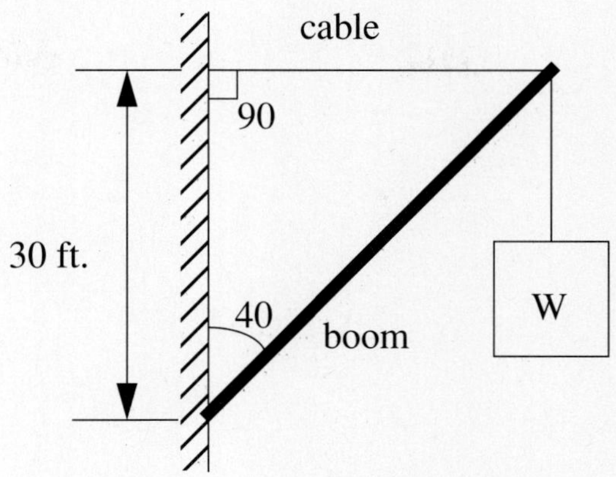

Figure D. Weight suspended by cable

6. Using Word, create the following figure:

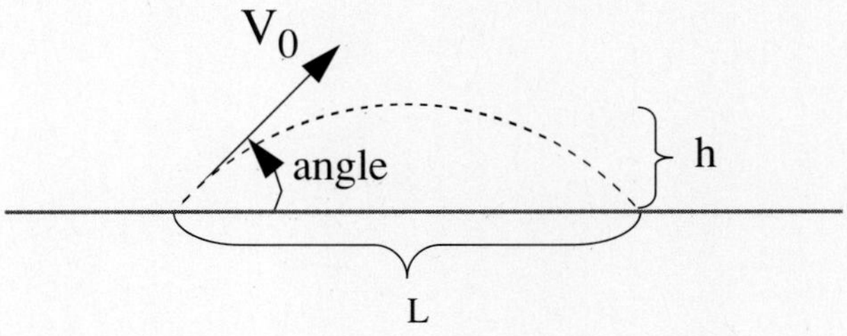

Figure E. Path of a projectile

7. Redesign your bedroom floor plan, and then write a paper discussing your new design and some of its advantages. Include at least one figure in your paper.
8. Select a topic, and then go to the library to search for articles on that topic. Write a one-page paper about the topic, and include at least one figure in your paper.
9. A number of well-publicized engineering failures have occurred over the years. Choose one of the failures, and write a paper about it. Include at least one figure in your paper.

7 Creating Design Specifications

Laser Technology

We are living in the Information Age. More information is readily available today than at any other time in history. You can retrieve information from satellites, newspapers, the Internet, and other sources with a personal computer and modem. As the amount of available information has increased, so has our need for data storage. The first floppy disks were 8 inches in diameter and very unstable. Today's compact discs (CDs) are smaller, more stable, and can store the same amount of data as 2,000 of the 8-inch disks. CDs use optical pick-up technology with lasers, which is a faster and more efficient way to access data than the magnetic pick-up technology used by floppies. Adopting new technologies to radically modify existing design solutions is characteristic of an ongoing design process. In this chapter, you learn about the fifth stage in the design process and about writing design specifications.

INTRODUCTION

In the fifth stage of the engineering design process, the solution is implemented. There is a building to be constructed, a road to be built, a product to be manufactured, or a system to be installed. Sometimes a product or project needs to be modified or improved. The engineer in charge of the implementation phase of the design process is not usually the same engineer who was responsible for the preliminary and design phases. For this reason, communication is vital. At this stage of the design process, engineers use a type of written communication called design specifications or design specs. This chapter discusses the final phase of the design process and explains how to format and create design specifications. You also learn about some Word features that you can use to create design specifications.

7-1 IMPLEMENTING AND TESTING THE SOLUTION

In the fifth stage of the design process, *design implementation*, engineers build or manufacture their solution. Typically, a different group of engineers carries out the implementation phase than those of the design phase. At a large company, different divisions are generally in charge of manufacturing the necessary parts and assembling the final product. At smaller firms, machining companies may be contracted to create the individual parts, and the assembly may be done at the original firm. In building and road construction projects, the design engineering firm usually hires a general contractor to implement the design. The general contractor often hires subcontractors who are each responsible for a portion of the entire project.

Because engineers who were not involved in the design are often responsible for the implementation of the project, clear communication is critical. At this stage, the documentation consists of at least two parts:

- A set of drawings that graphically illustrate the design solution
- A set of written specifications that supply any information not provided by the drawings

A third common type of documentation is a written contract between the engineers who carried out the design and the engineers responsible for the project implementation.

For example, suppose you are a chemical engineer responsible for the design and construction of an oil refinery. During the design process, several engineering teams design all the systems for the entire plant. A civil engineer designs the plant structure, a mechanical engineer designs the pumping and the heating and cooling systems, an electrical engineer identifies the power needs of the plant, and an environmental engineer develops the required pollution control mechanisms. As the chemical engineer, you are responsible for determining how the material flows through the plant, what temperature should be maintained at each stage of the chemical processing, and the different compounds to be extracted or added to the product during processing.

After the plant is designed, a general contractor is hired to build the facility. The contractor in turn hires subcontractors to construct the various systems within the plant. During all this construction, explicit written communication is supplied to the contractor and subcontractors in the form of *design specifications*. Specifications include the following:

- The size and capacity of all pumps used in the plant
- The material characteristics and forming techniques for the concrete and other materials in the plant
- The government codes, standards, and regulations that must be followed
- The terms and responsibilities for hiring subcontractors
- The soil compaction required for the foundation of buildings within the plant
- The type of cabling to be used within the wiring systems
- The required overall performance characteristics of the various plant systems

In the remainder of this chapter, you will learn how to write technical design specifications, and you will examine the various Word features that you can use to write specifications.

7-2 FORMAT OF DESIGN SPECIFICATIONS

There are two basic methods for writing design specifications. Using one method, you specify the techniques and materials to be used during the construction process. Using the other method, you specify the desired outcome and leave decisions regarding methods and materials to the contractor or manufacturer, as long as the performance requirements are achieved. In either case, you should use clear language and write in a simple style so that the contractor, the subcontractors, and the job supervisors understand exactly what you mean. Some large firms may have employees who are responsible for writing all design specifications. The specifications accompany the drawings, and these make up the entire design documentation. Because the specifications are written documents, they take precedence in any legal proceedings that may occur as a result of the building contract.

Design specifications should be written as if they were instructions to the contractor concerning what must be done. For example, you might say "The job site shall be maintained in a neat and orderly condition and kept free from waste materials during the entire construction period."

Design specifications assume that the contractor is performing the work. Therefore, the phrase "the contractor" is never the subject of the sentence; instead, the material or method is the subject. In the example, "the job site" is the subject of the sentence; it is understood that the contractor is responsible for keeping the job site clean.

The specifications are an extremely important aspect of the design. In general, design specifications should include the following elements:

- **General requirements.** These specifications include the nonlegal and nontechnical requirements for the job, such as the date for completion of the work.
- **Reference sources.** Materials used in construction projects must meet minimum standards developed by various governmental agencies.

When you specify a material to be used for a construction project, you include the specific government regulation that applies to the material so that the contractors know where to look for this information when purchasing or testing materials.

- **Material evaluation techniques.** These specifications describe in detail the procedures to be followed in the evaluation of materials and the expected performance requirements.
- **Manufacturers or sources for materials.** Sometimes you may want to specify a source for materials or products. For example, if you wanted to use a specific brand of pumps or other electro-mechanical components in the construction, you could include this information in the specifications.
- **Provisions for change.** During construction, there will be deviations from the original plans and specs. For example, as the oil refinery is constructed, some of the pipes may interfere with each other, requiring modification of the plant design. You may include provisions for tolerances and change as a part of the specifications.
- **Contractor's and subcontractor's limits.** These specifications delineate the responsibilities of the various contractors and subcontractors on a given job.
- **Insurance requirements.** Contractors and subcontractors are required to carry insurance covering workmen's compensation, fire damage, liability, and so on. You outline the provisions for this insurance and the minimum insurance standards in this portion of the design specifications.
- **Design criteria.** Use this type of specification when a drawing cannot adequately show all the design criteria. For example, you can specify the type of door finish to use or the number of coats of paint to apply.

The number and type of specifications you write for a given design solution vary depending on the complexity of the project; its function; and the governing federal, state, or local regulations.

7-3 CREATING DESIGN SPECIFICATIONS IN WORD

You can use all the Word features you have already learned about to create design specifications. Because specifications typically contain lists of criteria, in this chapter you first learn how to create bulleted lists. You also discover how to search for and replace text, and how to add special characters to a document.

Creating Bulleted Lists

When preparing engineering documentation, you have many opportunities to create *bulleted lists*—lists with a bullet (•) preceding each text item. Word allows you to insert and customize bullets. By default, Word places a bullet at the left margin of the page and moves the insertion point 0.25 inch to the right so that you can type the list item text. To begin a bulleted list, choose Bullets and Numbering from the Format menu. A dialog box like the one shown in Figure 7-1 appears.

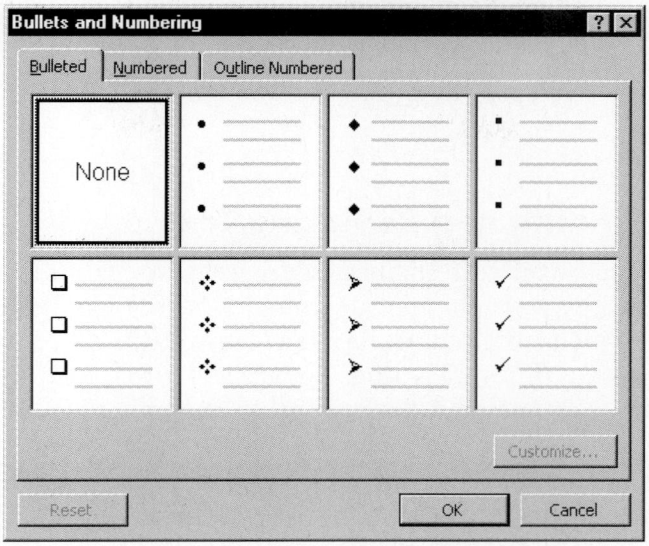

Figure 7-1
Bullets and Numbering dialog box

This dialog box permits you to select the type of bullet to be used. The available types of graphical bullets include small and large circles, diamonds, stars, arrows, boxes, and check marks. You can also use numbers, Roman numerals, or letters of the alphabet if you select the Numbered tab. A new bulleted item appears each time you start a new paragraph by pressing ENTER. When you are finished creating the bulleted or numbered list in your document, press ENTER one more time to create an extra bulleted list item. From the Bullets and Numbering dialog box choose None to turn this feature off and resume normal document editing.

Try It

In this exercise, you create a design specification that includes a bulleted list.

- Start Word. Type the following sentence:

  `The heating system for the facility shall meet the following performance standards:`

  Then press ENTER.
- Turn on the Bullets feature. (A bullet should appear at the left margin, and the insertion point will be located 0.25 inch to the right of the bullet.)
- Type the following list, pressing ENTER at the end of each list item:
  - `The heating plant shall be capable of providing an interior temperature of 70 F when the outside temperature is 0 F.`
  - `The heating system shall use No. 5 oil.`
  - `The heating system shall be a hot-water system.`
  - `The heating elements shall be baseboard radiation.`
  - `Controls such as thermostats and other safety devices shall be provided to regulate heat.`
- Press ENTER, and then remove the bullet from this new paragraph.
- Type the following sentence:

```
The components used in the heating system shall be
approved by the engineer prior to their installation.
```

- Save your work in a file named SPECS1.DOC.

Finding and Replacing Text

When you edit engineering documents, you may need to search for text strings within the document. Word lets you search for particular sequences of characters and lets you find text strings and replace them with a different string. To access Word's Find feature, choose Find from the Edit menu. A dialog box like the one shown in Figure 7-2 appears.

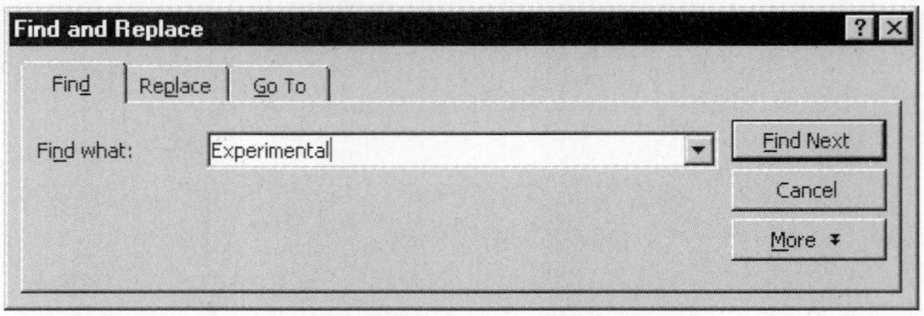

Figure 7-2
Find dialog box

Type the string of characters you wish to find in the Find What text box. In Figure 7-2, the word *Experimental* is the text string that Word will find. If you choose Find Next, Word searches forward through the document for the indicated text. If you select the option More, you can use other options found in an expanded version of the dialog box to customize your search. For example, if you select Match Case, the case (uppercase or lowercase) of the characters will be considered in the search. You can also specify that the search locates whole words only or that the font of the character string is considered.

You can replace one string of text with another by choosing Replace from the Edit menu (or by selecting the Replace tab in the Find and Replace dialog box). A dialog box like the one shown in Figure 7-3 appears.

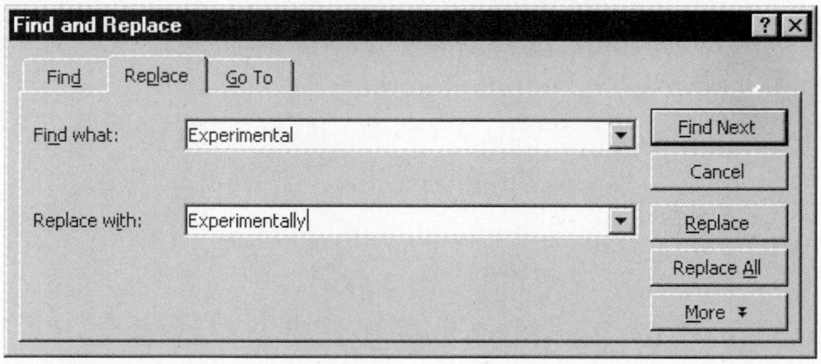

Figure 7-3
Replace dialog box

Note that this dialog box includes an additional text box for the replacement text string. In Figure 7-3, the word *Experimental* will be searched for and then replaced with the word *Experimentally.* If Find Next is selected from the dialog box, Word searches for the appropriate character string. You can then choose either Replace, to replace just this one occurrence of the text string, or Replace All, to replace all occurrences of the text string. As before, you can choose More to further customize your search.

Including Special Characters

Many times you need to include special characters (such as characters from the Greek alphabet) within your Word document. To include special characters within your document, choose Symbol from the Insert menu. A dialog box similar to the one shown in Figure 7-4 appears. From this dialog box, you can graphically select the character you wish to include in your document at the insertion point, and then choose Insert. Other symbol fonts are available by selecting them from the Font drop-down list box. When you are finished including special characters, you can Close this dialog box and resume editing your document.

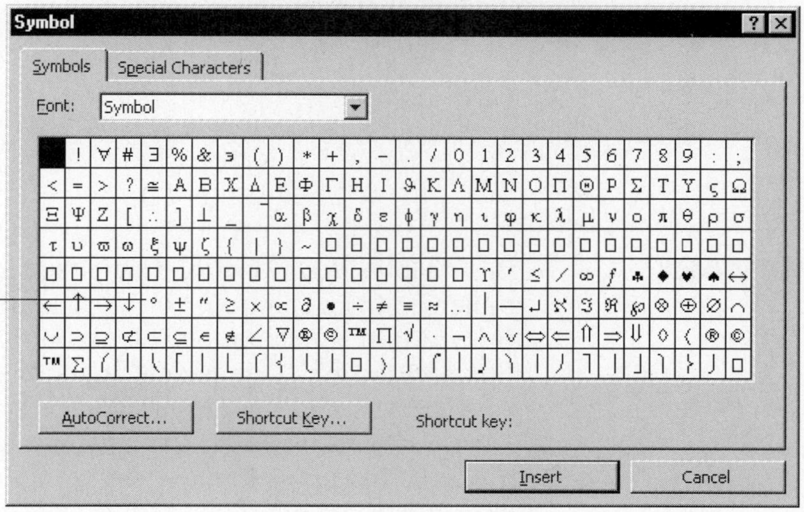

Figure 7-4
Symbol dialog box

Inserting Hard Page Breaks

By inserting *hard page breaks*, you can force the printer to start printing a new page at a specific location. For example, you may need to insert hard page breaks when you create bulleted lists to make sure that the entire list is located on one page. To insert a hard page break into your document, select Break from the Insert menu. From the dialog box that appears you can insert a page break at the insertion point location. The insertion point then moves to the top of the next page.

Try It

In this exercise, you insert the degree symbol (°) within your design specifications so that the temperatures read °F instead of just F.

- Open SPECS1.DOC if it is not already open.
- Place the insertion point at the beginning of the document containing the design specification you entered earlier.
- Use Word's Find and Replace feature to search for F. If you choose More and match the case during the search, you will find only an uppercase F.

- Use Word's special characters to replace F with °F within the Replace dialog box. The degree symbol is shown in Figure 7-4.
- Repeat the Find and Replace for the second occurrence of the F in the document.
- Insert a hard page break at the beginning of the paragraph that starts with "The components…."
- Obtain a printout of your document to turn in to your instructor.

7-4 USING SHORTCUTS

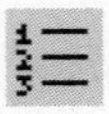
Creates a numbered list.

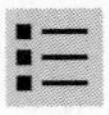
Creates a bulleted list.

SUMMARY

This chapter described the final stage of the engineering design process: implementing a solution. You learned how to create design specifications, the type of documentation you use in this step. You learned that engineering specifications supplement the information contained in the drawings for a project, and that design specifications must be written in clear and concise language so that the person implementing the design can readily grasp the intent of the designer. You also learned how to create bulleted lists, find and replace text strings, add special characters, and insert hard page breaks.

Key Words

bulleted lists
design implementation
design specifications
hard page breaks

Exercises

1. Write a specification about how to paint your room. Include a bulleted list that contains such information as the number of coats of paint, the application method, the color of the paint, and the type of paint (that is, enamel, latex, or semigloss).

2. Rewrite the following specification so that the language is clearer. Use the symbol σ_y in place of "yield strength" and σ_u in place of "ultimate strength" within the specifications:
   a. The contractor should use lightweight concrete with a minimum compressive strength of 3,000 psi for floor fill and slabs. You should use stone concrete for the foundations.
   b. The contractor should use reinforcing steel that is intermediate grade and that conforms to ASTM A615, 60,000 psi yield strength.
   c. You should use mesh reinforcing that is electric welded wire fabric conforming to ASTM A185 and A82 and that has an ultimate strength of 70,000 psi.
3. Almost all job sites require some site work, including the removal of trees, shrubs, and so on; demolition of existing structures and their foundations; removal of topsoil; disposal of excavated material; and installation of drainage systems. Write a specification that contains a numbered list detailing the site work required for a particular project.
4. Write a specification that contains a bulleted list about what you expect of someone cleaning your house.

Index